ELECTRICITY 3

ELECTRICITY

3

POWER GENERATION AND DELIVERY

NINTH EDITION

JEFF KELJIK

DELMAR
CENGAGE Learning

Australia • Brazil • Japan • Korea • Mexico • Singapore • Spain • United Kingdom • United States

Electricty 3: Power Generation and Delivery, Ninth Edition

Jeff Keljik

Vice President, Career and Professional Editorial: Dave Garza

Director of Learning Solutions: Sandy Clark

Senior Acquisitions Editor: John Fedor

Managing Editor: Larry Main

Senior Product Manager: Sharon Chambliss

Senior Editorial Assistant: Dawn Daugherty

Vice President, Career and Professional Marketing: Jennifer McAvey

Executive Marketing Manager: Deborah S. Yarnell

Senior Marketing Manager: Jimmy Stephens

Marketing Specialist: Mark Pierro

Production Director: Wendy Troeger

Production Manager: Stacy Masucci

Content Project Manager: Cheri Plasse

Art Director: Benj Gleeksman

Technology Project Manager: Chrstopher Catalina

Production Technology Analyst: Thomas Stover

For product information and technology assistance, contact us at
Cengage Learning Customer & Sales Support, 1-800-354-9706

For permission to use material from this text or product, submit all requests online at **www.cengage.com/permissions**
Further permissions questions can be emailed to
permissionrequest@cengage.com

Library of Congress Control Number: 2008925009

ISBN-13: 978-1-4354-0029-0

ISBN-10: 1-4354-0029-1

Delmar
Executive Woods
5 Maxwell Drive
Clifton Park, NY 12065
USA

Cengage Learning is a leading provider of customized learning solutions with office locations around the globe, including Singapore, the United Kingdom, Australia, Mexico, Brazil, and Japan. Locate your local office at **www.cengage.com/global**

Cengage Learning products are represented in Canada by Nelson Education, Ltd.

To learn more about Delmar, visit **www.cengage.com/delmar**

Purchase any of our products at your local college store or at our preferred online store **www.cengagebrain.com**

Notice to the Reader

Printed in the United States of America
3 4 5 6 7 15 14 13 12 11

CONTENTS

PREFACE

The ninth edition of *ELECTRICITY 3* has been updated to provide more topics and concepts to better reflect the current workplace. At the same time, the text has retained the features and style of previous editions that have made it so popular.

The text introduces the concepts of power generation and distribution. The material is broken down into short segments that concentrate on specific concepts or applications of particular types of equipment. The detailed explanations are written in easy-to-understand language and concisely present the needed information. Many illustrations and photographs help provide technical understanding and real-world references. This type of explanation and application better prepares the student to perform effectively on the job in installing, troubleshooting, repairing, and servicing electrical power generation and delivery.

The knowledge obtained in this book permits the student to progress further in the study of electrical systems. The study of electricity and the application of electrical products are continually changing. The electrical industry constantly introduces new and improved devices and material that lead to changes in installation and operation of equipment. Electrical codes also change to reflect industry needs. It is essential that students continue to learn and update their knowledge of current procedures and practices.

The text is easy to read and the units have been grouped by general subject areas. Summaries in each unit restate the most important topics of the unit. Summaries of unit groupings provide reviews of topical areas.

Each unit begins with the learning objectives. An Achievement Review at the end of each unit provides an opportunity for readers to check their understanding of the material in small increments before proceeding. The problems in the text sometimes require the use of simple algebra. Students should be familiar with the math before trying to solve the equations. It is also essential that readers have a basic understanding of the fundamentals of electrical circuits and basic electrical concepts.

It is recommended that the most recent edition of the *National Electrical Code*® (published by the National Fire Protection Association [NFPA]) be available for reference and use as the learner uses this text. Application of state and local codes and regulations should also be consulted when making actual installations.

Features of the ninth edition include the addition of a unit on alternative power sources which include wind-powered generation, hydrogen fuel cells, microturbine generation, and solar photovoltaic power production. These alternative forms of power supplies are described and then linked to the *National Electrical Code*®, where applicable, to enhance the understanding of where and how these sources are used. An alternate form for a UPS system is described, to be used for short duration operation in place of battery backup systems.

Other features include

- Organization of topics into related topics and associated concepts
- Updated photos and artwork to reflect current equipment and practices

- Content updated to the most recent electrical code
- Additional information on generation systems
- Coverage of transformers and connections
- Summaries and Achievement Reviews at the end of each unit

An instructor's guide for *ELECTRICITY 3* is available. The guide includes the answers to the Achievement Reviews and Summary Reviews and additional test questions. Instructors may use these questions to devise tests to evaluate student learning.

ABOUT THE AUTHOR

Jeff Keljik has been teaching at Dunwoody Institute in Minneapolis for more than 30 years, where he was the head of electrical programs for more than 16 years. He is a licensed master and journeyman electrician and has a bachelor's degree in business communication. He currently teaches classes for corporate clients locally and nationally. He also manages the electrical construction and maintenance projects for the college campus buildings at Dunwoody Institute. He has worked as an electrician and as a consultant on international training projects in the electrical industry.

In addition to his teaching and administrative positions, Mr. Keljik serves the North Central Electrical League as Education Chair and as a member of the Board of Directors. He also serves as an advisor on the Education committee member for the Minnesota Electrical Association (MEA). He has written several texts on motor and motor control systems and on power generation and distribution, including *Electricity 4* and *Electric Motors and Motor Controls*.

ACKNOWLEDGMENTS

Grateful acknowledgment is given to the following individuals for their contributions to this edition of *Electricity 3:*

David Adams
Niagara County Community College
Sanborn, NY

Joseph T. Brown
Tri County Technical College
Pendleton, SC

Phillip Serina
Kaplan Career Institute
Brooklyn, Ohio

DEDICATION

I would like to dedicate this ninth edition to my daughter, Katherine. Her love of learning encourages me to continue to write texts that help others learn.

— Jeff Keljik

ELECTRICAL TRADES

The Delmar series of instructional material for the basic electrical trades consists of the texts, text-workbooks, laboratory manuals, and related information workbooks listed below. Each text features basic theory with practical applications and student involvement in hands-on activities.

ELECTRICITY 1
ELECTRICITY 2
ELECTRICITY 3
ELECTRICITY 4
ELECTRIC MOTOR CONTROL
ELECTRIC MOTOR CONTROL
LABORATORY MANUAL
INDUSTRIAL MOTOR CONTROL
ALTERNATING CURRENT
FUNDAMENTALS

DIRECT CURRENT FUNDAMENTALS
ELECTRICAL WIRING—
RESIDENTIAL
ELECTRICAL WIRING—
COMMERCIAL
ELECTRICAL WIRING—
INDUSTRIAL
PRACTICAL PROBLEMS
IN MATHEMATICS
FOR ELECTRICIANS

Equations based on Ohm's Law.

P = Power in watts
I = Intensity of current in amperes
R = Resistance in ohms
E = Electromotive force in volts

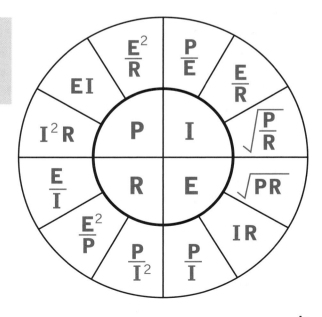

INTRODUCTION

ELECTRICAL SAFETY

Working on electrical systems can be dangerous. If you are working on live exposed conductive paths and the voltage is over 50 volts to ground, the potential for electrocution exists. The potential for arc burns or explosions also exists when working on live electrical systems. The United States Occupational Safety and Health Administration (OSHA) has specific guidelines to follow when working on live equipment. The OSHA documents that primarily affect practicing electrical workers are found in the OSHA Code of Federal Regulations (CFR) 29–1910. The regulations have many subparts that specify requirements whether you are doing electrical construction or electrical maintenance. In addition to OSHA, the *National Electrical Code*® refers to the *National Fire Protection Association (NFPA) document 70E*. This NFPA document is developed by the organization that develops the *National Electrical Code*® (*NEC*®).

NFPA 70E correlates the requirements of OSHA and interprets them for use by electrical personnel. The *70E* document is considered a consensus standard, which means governing agencies must adopt or consent to use it as a standard. *NFPA 70E* and OSHA standards are very similar in intent; *70E* interprets the standards in application-related terms. *NFPA 70E* provides guidelines on safe work practices, information on how to determine arc flash boundaries, and guides on what personal protective equipment (PPE) to use in various situations.

YOUR SAFETY IS YOUR RESPONSIBILITY. While agencies and organizations try to provide guidelines and practices to follow, it is up to you to be as safe as possible in your work responsibilities. If at all possible, disconnect power from all the circuits with which you will come in contact. You must check for the presence of electrical energy and then you must lock out the system to prevent accidental re-energization. The term "Lock-Out/Tag-Out (LOTO)" is often used to ensure that you place your personal lock on a disconnection point and then tag it with your name to identify that you are working on the circuit. No one should remove your lock without your knowledge and consent.

There are cases when you will work on systems that are energized. OSHA requires that you have a thorough knowledge of the precautions and procedures to follow. You must have documented training and verification that you are knowledgeable of the dangers involved, you know the precautionary techniques to use, you are aware of the electrocution hazards, and you understand the hazards of arc flash. You must wear appropriate PPE that has been determined adequate to protect you from the inherent hazards. The PPE includes specifications for the proper clothing and hand and face protection as well as properly rated tools. You must use meters and test gear that accurately test for conditions that are present. You must have a plan to follow if an accident occurs and you need help. There are rules for the number of people who need to be present and there are rules for entering a confined space. Be aware of underground installations and the need for fresh air to breathe. If the environment is hazardous, for example if gas vapors, dust, or fiber material are present, take further precautions as required.

Arc flash is a danger that is present as you work on live equipment, even if you do not physically touch the conductors with your hands. Electrocution or severe shock may result in contact with the live conductors, but the arc causes damage without contact. The arc that can occur between live parts, or live parts and a grounded surface, releases a great deal of thermal energy. The arc produces magnetic energy and air pressure energy known as the blast. This energy can knock you down or knock the air from your lungs. This same blast creates noise that can damage your hearing. The light produced from the arc may damage your eyesight. The amount of energy that may affect you depends on the type of fault, the amount of short-circuit current available at the arc, the voltage available, and your proximity to the fault. The arc heat energy is rated in calories and the protective clothing you wear must meet or exceed the number of calories present in a possible arc. Your hands must be protected from accidental contact by using gloves that meet or exceed the voltage rating of the circuit. See the figure showing voltage-rated gloves with leather protection. Your tools must be marked with the voltage rating for the circuit. Voltage-rated tools are pictured in the accompanying figure.

Rubber gloves rated for circuit voltage and leather gloves to protect the insulating qualities of the rubber.

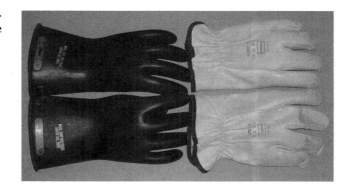

Tools used when working with live electrical circuits need to be electrically insulated and marked with the maximum circuit voltage.

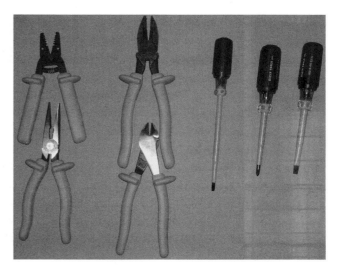

OPERATING PRINCIPLES OF DC GENERATORS

OBJECTIVES

After studying this unit, the student should be able to

- state the function of a direct current (DC) generator.
- list the major components of a generator.
- describe the difference between a separately excited and a self-excited generator.
- explain how the output voltage of a generator can be varied.

A *DC generator* changes mechanical energy into electrical energy. It furnishes electrical energy only when driven at a definite speed by some form of prime mover, such as a diesel engine or a steam turbine.

DC generators are used principally in electrical systems for mobile equipment. They are also used in power plants supplying DC power for factories and in certain railway systems. DC power is used extensively in communication systems and for battery charging and electroplating operations. The generation of electromotive force (EMF) is described in detail in *Electricity 1*.

In a DC generator, the output voltage is a DC voltage, even though alternating current (AC) is generated in the generator coils. In other words, one of the output terminals will always be negative (–) and the other output terminal will always be positive (+). The external circuit is the load connected to the generator. It receives current from the negative terminal of the generator and returns it to the positive terminal of the generator.

DC GENERATOR COMPONENTS

The essential parts of a DC generator are shown in Figures 1-1 and 1-2. The member that spins is called the *rotor*. The rotor is a cylindrical, laminated iron core that is mechanically coupled to the drive shaft of the generator. An armature winding is embedded in the slots on the surface of the rotor. The armature windings have voltage induced into them as the rotor spins past the field poles.

The *armature* is the component of the generator that has the voltage (EMF) induced into it. The internal circuit of the generator supplies electrons to the negative output terminal of the generator and creates a deficit of electrons at the positive terminal of the

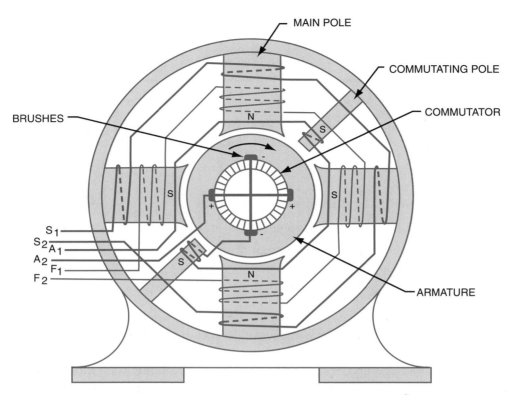

Figure 1-1 Compound generator fields with commutating poles.

generator. The electrons are drawn from the load back to the positive terminal of the generator. The armature is composed of coils of copper conductors that are terminated at the *commutator* segments. The commutator segments and the connection to the sliding carbon brushes provide the mechanical rectification needed to convert the induced alternating-current EMF within the armature to the direct-current EMF at the carbon brush connections to the load.

A *commutator* consists of a series of copper segments that are insulated from one another and the shaft. The commutator turns with the shaft and the armature windings. The commutator is used to change the AC voltage induced in the armature windings to DC voltage at the generator output terminals. Carbon brushes pressing against the commutator segments connect the current to the external load circuit.

The armature windings generate voltage by cutting a magnetic field as the armature rotates. This magnetic field is established by electromagnets mounted around the periphery of the generator. The electromagnets, called *field poles,* are arranged in a definite sequence of magnetic polarity; that is, each pole has a magnetic polarity opposite to that of the field poles adjacent to it. Electrical current for the generator field circuit is usually obtained from the generator itself.

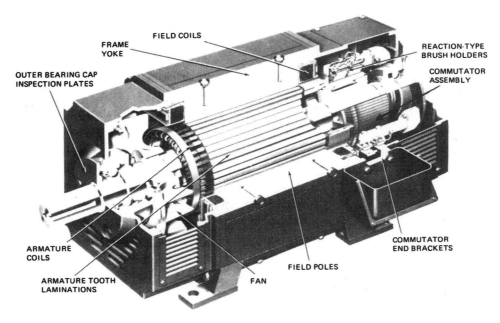

FIELD COILS

FRAME
YOKE

REACTION-TYPE
BRUSH HOLDERS

COMMUTATOR
ASSEMBLY

OUTER BEARING CAP
INSPECTION PLATES

COMMUTATOR
END BRACKETS

ARMATURE
COILS

ARMATURE TOOTH
LAMINATIONS

FAN

FIELD POLES

Figure 1-2 Cutaway view of a direct-current generator. (*Courtesy of Reliance Electric*)

When a generator supplies a current to a load, the current passes through the armature conductors too. As this current passes through the armature conductors, it creates a magnetic field around the armature windings. This magnetic field creates a magnetic influence in the rotor—a counter torque—that opposes the direction in which the rotor is being driven. This is called the "motor effect" of the generator. Figure 1-3 shows that the magnetic field created opposes the driving force. The larger the load current through the generator, the larger the counter torque produced. This is the reason that a generator takes more mechanical power to drive a heavy current load than a light current load.

Armature Reaction

The armature field flux also reacts against the main field flux and tends to distort it. One result of this undesirable condition, known as *armature reaction,* is excessive sparking at the brushes on the commutator. To counteract this effect, commutating poles are often inserted between the main field poles, as shown in Figure 1-1. These commutating poles, also called *interpoles,* are energized by windings placed in series with the output (load) circuit of the generator. Because of this arrangement, armature reaction, which tends to increase with load current, is counteracted by the effects of the load current passing through the interpoles.

Armature reaction, appearing as excessive brush sparking under load, also can be partially corrected by shifting the brushes from their neutral position in the direction of rotation.

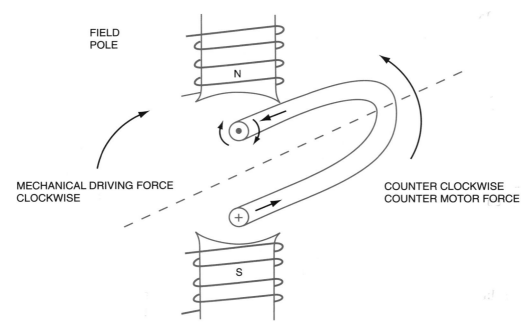

FIELD
POLE

N

MECHANICAL DRIVING FORCE
CLOCKWISE

COUNTER CLOCKWISE
COUNTER MOTOR FORCE

S

Figure 1-3 "Motor action" opposing generator driving force.

Large DC generators have the brushes assembled so that they can be shifted to the position of minimum sparking. When the brushes are not movable, the generator manufacturer inserts other design features to minimize the effects of armature reaction.

Because the magnetic field is distorted by the magnetic action of the armature current in the rotating armature, the neutral plane of the motor is shifted and the brushes are no longer at the exact center of the magnetic plane. This reaction to the armature current causes the sparking of the brushes. If the load current were always constant, the brushes could be shifted in the direction of rotation until the sparking stops. If the load current is constantly changing, interpole windings are inserted and connected in series with the load to counteract the shifted magnetic field and, in effect, bend it back to its original physical position. When the DC generator is large enough, interpoles are insufficient to counteract the effects of the shift in the neutral plane. Compensating windings are then wound into the pole faces and connected in series with the output current to help move the neutral magnetic plane back to the original physical location.

Another effect of armature reaction is the reduction of the output voltage. As the magnetic field of the rotor (armature) increases because of increased load current, the magnetic field of the rotor tends to cancel some of the magnetic field strength of the main field. This canceling effect tends to decrease the voltage output of the generator.

Brush Polarity

The output terminals of a generator, as with other DC power units, have electrical polarity. In the case of generators, the term *brush polarity* is used to distinguish between the *electrical polarity* of the brushes, either positive or negative, and the *magnetic polarity,* either north or south, of the field poles.

Brush polarity markings are often omitted, but the electrician can easily determine electrical polarity by connecting a voltmeter across the output leads of the generator. Many automotive and aircraft generators are constructed with either the positive or negative brushes grounded to the frame of the generator. It is very important to maintain the polarity as specified by the manufacturer. Additional information on brush polarity will be given after the effects of residual magnetism in the field circuit are considered.

Field Supply

The magnetic field of a generator is established by a set of electromagnets (field poles). The current required by the field circuit may be supplied from a separate DC supply. If this is the case, the generator is said to have a *separately excited field.* The majority of generators, however, are *self excited* and the current for the field is supplied by the generator itself.

Figure 1-4 illustrates a separately excited DC generator with the field circuit supplied from batteries. A self-excited shunt generator is shown in Figure 1-5. Note that the field circuit is connected in parallel with the armature and that a small part of the generator output is diverted to the field circuit to "excite" or energize the field poles.

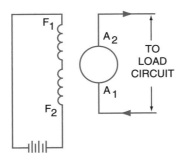

Figure 1-4 Separate excitation.

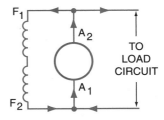

Figure 1-5 Self-excitation.

OUTPUT VOLTAGE CONTROL

Because the induced voltage into the armature depends on the rate at which the magnetic lines of force are cut per second, it is possible to vary the output voltage by controlling either the speed of the prime mover or the strength of the magnetic field. In all but a few instances, the output voltage is controlled by varying the field current in the field circuit.

The flux density in the magnetic field poles depends on the field current. As a result, the field strength and voltage output of the generator continue to increase with an increase of field current until magnetic saturation of the field poles occurs. Any additional increase in voltage output after this point must be obtained by an increase in speed of the rotating armature.

GENERATOR RATINGS

Generator ratings as specified by the manufacturer shall be found on the nameplate of the machine. The manufacturer generally specifies the kilowatt output, current, terminal voltage, and speed of the generator. For large generators, the ambient temperature is also given. See *National Electrical Code®, Article 445.11.*

ROTATION

A separately excited generator develops voltage for either direction of rotation. This is not true, however, for self-excited units; they develop voltage in one direction only. (See explanation in Unit 3.) The standard direction of rotation for DC generators is clockwise when looking at the end of the generator opposite the drive shaft (this is usually the commutator end).

REGULATION

The voltage regulation of a generator is one of its important characteristics. Different types of generators have different voltage regulation characteristics.

Figure 1-6 shows the action of the voltage at the terminals of a generator for different values of the load current. The drop in terminal voltage is caused by the loss in

**Figure 1-6
Voltage regulation
graphs.**

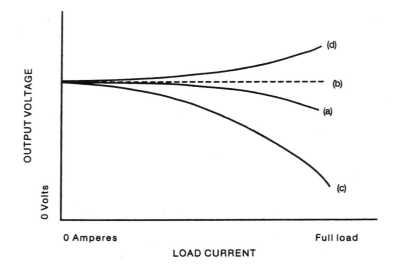

voltage due to (1) the internal resistance of the armature circuit including the brush contacts, and (2) armature reaction. The curve at (a) is the normal curve for a shunt generator. An ideal condition is shown in (b) where the voltage remains constant with load current. Curve (c) illustrates a generator with very poor regulation in that the output voltage drops off considerably as the load current increases. A rising characteristic, curve (d), is obtained by using a cumulative compound-wound generator (see Unit 4).

SUMMARY

DC generators are used to provide *direct current* to specific loads. The armature, mounted on the rotor, is driven through the magnetic poles developed by the electromagnetic fields. AC voltage is actually induced into the armature, and then mechanically rectified by the commutator and brush assembly. Brush polarity is established in a self-excited generator by the magnetic polarity of the poles and the direction of rotation. Voltage output is determined by the speed of rotation and the strength of the magnetic field.

ACHIEVEMENT REVIEW

Select the correct answer for each of the following statements and place the corresponding letter in the space provided.

1. A generator   _B_
    a. changes electrical energy to mechanical energy.
    b. changes mechanical energy to electrical energy.
    c. is always self excited.
    d. is always separately excited.

2. One of the following is not essential in generating a DC voltage: _C_
    a. a magnetic field   c. slip rings
    b. a conductor   d. relative motion

3. Commutating poles are   _B_
    a. fastened to the center of the commutator.
    b. located midway between the main poles.
    c. secondary poles induced by cross magnetizing the armature.
    d. used to regulate the voltage at the armature.

4. The winding on an interpole is   _C_
    a. made of many turns of fine wire.
    b. wound in a direction opposite to that of the armature winding.
    c. connected in series with the armature load.
    d. connected across the generator terminals.

5. Generator terminals A_1 and A_2 are terminals that   *A*
    a. connect the armature only.
    b. connect the shunt field in series with the armature.
    c. connect the series field to the armature.
    d. have the armature in parallel with the commutating poles.

6. To raise generator voltage, the   *A*
    a. field current should be increased.
    b. field current should be decreased.
    c. speed should be decreased.
    d. brushes should be shifted forward.

7. Generator voltage output control is usually accomplished by   *A* *B*
    a. varying the speed.
    b. a rheostat in the field.
    c. increasing the flux.
    d. decreasing the flux.

8. In Figure 1-6, the normal voltage regulation for a shunt   *A*
    generator is at
    a. curve (a).   c. curve (c).
    b. a broken line.   d. curve (d).

2

THE SEPARATELY EXCITED
DC GENERATOR

OBJECTIVES

After studying this unit, the student should be able to use industry standards to

- explain the relationship of field current, field flux, and output voltage for a separately excited DC generator.
- describe the effects on the brush polarity of reversing the armature rotation and the field current.
- define residual flux and residual voltage.
- draw and explain the basic circuit.
- connect the generator.

The separately excited DC generator has few commercial applications, but knowledge of its operations is an excellent background for understanding other types of generators.

Using a separate source of DC power, S_1 is closed as in Figure 2-1. As DC current flows through the coil of wire wound around the iron core, a magnetic field is produced. The amount of *field current* is controlled by the resistance of the field winding and the variable resistor, known as the *field rheostat*. By adjusting the field current, the strength of the magnetic field is controlled. The *field flux* or magnetic strength of the magnetic poles is increased as the field current is increased until *magnetic saturation* occurs. Saturation of the magnetic field means that no more magnetic flux can be produced even with an increase in field current. The magnetic polarity of the field poles is controlled by the direction of the DC field current and the direction of the wound coils.

The output voltage of the generator is developed as an induced voltage in the armature conductors. This induced voltage appears at the brushes and the generator output terminals designated as A_1 and A_2 in Figure 2-1.

To more fully understand the direction of induced voltage, use the left-hand rule for generation, shown in Figure 6-2 in Unit 6. In this figure, the thumb represents the thrust or the direction of the armature movement. The first finger represents the flux direction of the main magnetic field, and the center finger represents the current flow resulting in

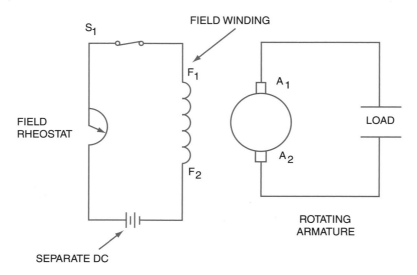

Figure 2-1　Separately excited DC generator connections.

the armature conductors. Note that by changing the direction of rotation or the direction of the main field flux, the direction of the resultant current flow is reversed.

The output voltage is *directly proportional* to the speed of the rotation and the strength of the magnetic field. As the speed of the rotor is increased, the output voltage will also increase. There is, however, a limit to the safe operating speed of the rotor before physical damage occurs. Likewise, the output voltage can be controlled up to a point by adjusting the field current.

BRUSH POLARITY

When the armature is driven in either direction, an electrical polarity is established at the generator output terminals and at the brushes. If the machine is stopped and then driven in the opposite direction, the field flux is cut in the opposite direction and the brush polarity changes, as shown in Figure 2-2(A).

If the direction of rotation is not changed and the field current is reversed, the same effect is obtained; that is, if the armature conductors maintain a rotation in one direction and field flux is established in the opposite direction, then the brush polarity also changes, as shown in Figure 2-2(B).

As a result, the brush polarity in a separately excited generator can be changed by reversing the rotation of the armature or the direction of the field current. However, if both the armature direction and the direction of the field current change, the brush polarity remains the same (unchanged). *polarity of Field poles*
direction of rotation of the armature

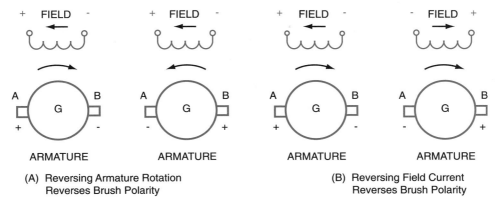

(A) Reversing Armature Rotation
Reverses Brush Polarity

(B) Reversing Field Current
Reverses Brush Polarity

Figure 2-2 Factors affecting brush polarity.

OUTPUT VOLTAGE

The magnitude of the voltage depends on the rate at which the flux is cut. In a separately excited generator, an output voltage increase is proportional to an increase in the armature speed. The upper limit of the voltage is determined by the permissible mechanical speed and the insulation qualities of the armature and the commutator.

The output voltage of a separately excited generator can be varied by adjusting the speed of the armature rotation or the field current. A change in speed always results in a corresponding change in output voltage. An increase in field current increases the output voltage only if the field poles are not saturated. Field control of the output voltage is accomplished by varying the total resistance of the field circuit with a field rheostat, as shown in Figure 2-1. Increasing the resistance reduces the field current and reduces output voltage. Decreasing the resistance increases the output voltage.

RESIDUAL VOLTAGE

If the field circuit is opened at S_1 (Figure 2-1) the field current becomes zero. A small amount of magnetic flux called *residual flux* remains, which is caused by residual magnetism left in the field poles. The small voltage generated when the armature cuts this flux is called *residual voltage*. Brush polarity remains the same when the field current is zero because the residual flux has the same direction as the original flux. If the armature is rotated in the opposite direction, the same residual voltage is obtained at the same speed, but the brush polarity reverses. If the field circuit is closed momentarily with the battery connections reversed, the residual flux reverses and the brush polarity reverses.

SUMMARY

Generation of DC voltage depends on three factors: a magnetic field, motion, and conductors. Separately excited generators use a separate DC voltage to control the source of field excitation. By increasing the field current, field flux can be increased. By controlling the direction of field current through the coils, the magnetic polarity is established.

Output voltage level is controlled by the speed of the rotating armature and strength of the magnetic field. The polarity of the output voltage is controlled by the direction of rotation of the armature and the direction of the magnetic field.

ACHIEVEMENT REVIEW

A. Select the correct answer for each of the following statements and place the corresponding letter in the space provided.

1. A separately excited DC generator has the field connected   _C_
    a. across the armature.
    b. in series with the armature.
    c. to an external circuit.
    d. none of the above

2. F_1 and F_2 generator terminals are   _BA_
    a. shunt field leads.
    b. series field leads.
    c. armature leads.
    d. commutating pole leads.

3. The voltage of a separately excited DC generator   _QA_
    may be increased by
    a. increasing the speed of rotation of the armature.
    b. decreasing the magnetic flux.
    c. both a and b
    d. neither a nor b

4. The function of brushes on a generator is to   _A_
    a. carry the current to the external circuit.
    b. prevent sparking.
    c. keep the commutator clean.
    d. reverse the connections to the armature to provide DC.

5. Electrical polarity at the brushes may be changed by   _C_
    a. reversing the rotation of the armature.
    b. reversing the direction of the field current.
    c. either a or b
    d. neither a nor b

B. Select the correct answer to questions 6 through 9 from the following list and
    write it in the space provided.

power source	only one	decrease
always	strength of the field flux	sometimes
speed of the armature	either	increase

6. In addition to armature rotation, the output voltage varies with the ___Flux___.

7. One factor limiting an increase in output voltage is the ___Speed of armature___.

8. A change in the speed of rotation of the armature___always___ results in a
    change in the output voltage.

9. If the field poles are saturated, an increase in the field current does not cause a(n)
    ___increase___ in the output voltage.

3

THE SELF-EXCITED SHUNT GENERATOR

OBJECTIVES

After studying this unit, the student should be able to

- identify a self-excited shunt generator from a circuit diagram.
- describe how voltage buildup occurs for this type of generator.
- list the causes for a failure of the voltage to build up.
- describe three methods that can be used to renew residual magnetism.
- define voltage control and voltage regulation.
- draw the basic circuit.
- connect the generator.

Most DC generators of the shunt type are self-excited. A generator is called a *shunt generator* when its field circuit is connected in parallel with the armature and load. Within the field circuit itself, a four-pole winding may be connected in series, parallel, or series-parallel. The circuit arrangement of the field windings does not affect the classification of the generator because the field windings as a group are connected in parallel with the armature and load.

VOLTAGE BUILDUP

Figure 3-1 shows the schematic diagram of a self-excited shunt generator. Voltage control is obtained with a field rheostat. Unlike the separately excited generator, there is no current in the field circuit when the armature is motionless. Because a small amount of residual magnetism is present in the field poles, a weak residual voltage is induced in the armature as soon as the armature is rotated. This residual voltage produces a weak current in the field circuit. If this current is in the proper direction, an increase in magnetic strength occurs with a corresponding increase in voltage output. The increased voltage output, in turn, increases the field current and the field flux, which again increases the voltage output. As a result of this action, the output voltage builds up, increasing field current to the field poles. After the poles are established, the voltage remains at a constant level, unless the speed of the armature rotation is changed.

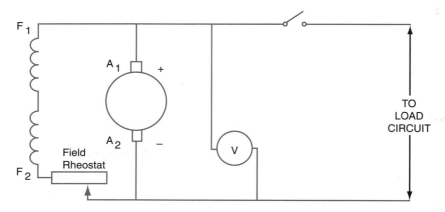

Figure 3-1 Self-excited shunt generator.

If the direction of armature rotation is reversed at start-up, the brush polarity also is reversed. The residual voltage now produces a field current that weakens the residual magnetism and the generator voltage fails to build up. Therefore, a self-excited machine develops its operating voltage for one direction of armature rotation only. The load switch for the generator may be closed when the desired voltage is reached.

LOSS AND RENEWAL OF RESIDUAL MAGNETISM

A shunt generator may not develop its rated operating voltage if the residual magnetism is lost. The residual flux may be renewed by momentarily connecting a low-voltage DC source across the field circuit. Several methods can be used to renew the residual magnetism. This process is called "flashing the field."

Method 1

1. Disconnect the field circuit leads from the brushes.
2. Momentarily connect a storage battery or low-voltage DC source to the field circuit leads. To maintain the desired brush polarity, connect the positive terminal of the battery to the field lead normally attached to the positive generator brush.

Method 2

1. If it is inconvenient to detach the field leads and the brush assembly can be reached, lift either the positive or the negative brush and insert a piece of heavy, dry paper between the brush and the commutator segments.
2. Momentarily connect a battery to the output leads. With the brush lifted, current passes through the field circuit only. (To maintain the original brush polarity, connect the positive terminal of the battery to the positive generator output terminal.)
3. Remove the paper under the brush before restarting the generator.

Method 3

1. If it can be done readily, disconnect the generator from its prime mover.
2. Restore the residual field by momentarily connecting a battery to the generator output leads. Because the field circuit is connected across the output leads, the current renews the magnetic field.

Caution: If the armature is not free to rotate, damage to the armature assembly may occur.

When the battery voltage is high enough in Method 3, the generator armature rotates as a motor. The rotation produced does not contribute to restoring the residual flux. However, this effect, called *motorizing,* is useful because it is a rough check of the overall generator operation. That is, the armature should rotate freely if the voltage applied is a sizable fraction of the rated output voltage, with the direction of armature rotation opposite to the proper direction of rotation for a generator. Be sure to use a reduced voltage for large motors.

Brush Polarity

To maintain the original brush polarity when renewing the residual magnetism, the electrical polarities of the output leads and the exciting battery must be matched. In other words, the positive terminal of the battery must be connected to the positive output terminal of the generator, and the negative battery terminal must be connected to the negative generator terminal.

The motorizing test should never be used for restoring residual flux if the generator armature is mechanically engaged to the prime mover and cannot rotate freely. A strong current through the motionless armature sets up a powerful magnetic field on the armature core. This magnetic field may overpower and reverse the main field flux, causing a reversal of the brush polarity when the generator is restarted. If there is any doubt as to whether the armature can be disconnected completely from the prime mover, it is preferable to isolate and energize the field circuit only, either by lifting the brushes or disconnecting the field leads.

CRITICAL FIELD RESISTANCE

A shunt generator may fail to reach its operating voltage even though its residual magnetic field is satisfactory. This failure may be due to excessive resistance in the field circuit. Any generator has *critical field resistance*. The presence of resistance in the field circuit in excess of this critical value causes the generator to fail to build up to its rated operating voltage.

Because field rheostats are used to control the voltage output at rated speed, it is important to reduce the resistance of the field rheostat to a minimum value before investigating other possible faults in the event of failure to develop rated voltage. Be sure a complete circuit exists from the field through the field resistance.

BRUSH CONTACT RESISTANCE

Contact resistance at the brushes is another reason for the failure of the generator to develop its operating voltage. Because the field circuit is *completed* through the armature, any resistance introduced at this point is effectively in the field circuit. Additional pressure applied to the brushes may indicate trouble from this source.

Improper connection of the field circuit leads at the brushes is also a cause of failure to build up rated voltage. An improper connection can be discovered by reversing these leads.

If there is insufficient pressure of the brushes riding on the commutator, the electrical connection is poor and the increased resistance causes too low a voltage to be available to the main field. Increasing pressure may alleviate such a high-resistance connection. To check for proper contact pressure, use a simple resistance test. With the power off, measure the resistance from the rotor segment to the associated brush pigtail with normal pressure applied. The reading should be very low ohms. If increasing pressure of the brush to the commutator significantly reduces the ohm value, then brush pressure may be a problem. Either replace the brushes with brushes that are not worn down or increase the spring pressure behind the brush.

ROTATION

When a DC shunt generator is used in special applications, it may be necessary for the armature to rotate in a direction opposite to that specified by the manufacturer. To develop voltage buildup in these instances, the field circuit leads at the brushes must be reversed.

RATINGS

Shunt generators are rated for speed, voltage, and current. Generators used in aircraft and automobiles operate through a wide range of speeds, but must maintain a constant load voltage. Voltage regulators that automatically change field resistance are used.

Generators designed for operation at a constant rated speed must not be operated above this value, unless the field circuit is protected from the effects of excessive current by current-limiting devices.

OUTPUT VOLTAGE CONTROL

Field rheostats are used to control the voltage output of shunt generators. At a given speed, the rheostat can be used only to bring the output voltage to values below the rated voltage obtainable without a field control. Values above the normal rated voltage can be obtained only by operating the generator above normal speed.

VOLTAGE REGULATION

The terms "voltage regulation" and "voltage control" are often confused. *Voltage control* refers to *intentional* changes in the terminal voltage made by manual or automatic regulating equipment, such as a field rheostat.

Voltage regulation refers to *automatic* changes in the terminal voltage due to reactions *within* the generator as the load current changes.

Voltage regulation is defined as the percent difference between the voltage output when there is no electrical load (E@NL) and the terminal voltage is at full-rated current capacity (E@FL). The formula used to determine the percent of voltage regulation follows:

$$\frac{E@NL - E@FL}{E@FL} \times 100 = \% \text{ regulation}$$

For example, it is inherent in the design of a shunt generator for the output voltage to fall off as the load increases if there is no change in field resistance. (See Unit 4.) If the drop is severe, the generator is said to have poor voltage regulation.

SUMMARY

A self-excited shunt generator has the field coils and the field rheostat shunted across the armature connections. If residual magnetism is left in the field iron, then spinning the armature will produce residual voltage. This residual voltage is normally enough to begin the generation process. If there is not enough residual voltage, then the residual magnetism must be reestablished. Self-excited generators must have the field polarity correctly established and the armature spinning in the proper direction to develop output voltage. The output voltage can be controlled by adding or removing resistance to the shunt field circuit.

ACHIEVEMENT REVIEW

Select the correct answer for each of the following statements and place the corresponding letter in the space provided.

1. Most DC generators are
   a. self-excited.
   b. excited by storage batteries.
   c. series wound.
   d. excited separately.

2. The field coils of a shunt generator are always connected
   a. in parallel with a rheostat.
   b. in parallel with each other.
   c. in series with each other.
   d. across the line.

3. The voltage of a shunt generator is built up by   ~~B~~ C
   a. permanent magnetism.
   b. proper operation of the field rheostat.
   c. residual magnetism.
   d. increasing the speed.

4. The field windings of a shunt generator must have   ~~B~~ D
   a. full line current applied.
   b. comparatively low resistance.
   c. 1 ohm of resistance per volt.
   d. comparatively high resistance.

5. Cutting resistance out of a shunt field circuit   D
   a. cuts down the magnetic flux.
   b. decreases the terminal voltage.
   c. increases the load.
   d. increases the terminal output voltage.

6. Failure of a DC shunt generator to build up to its   ~~A~~ F
   rated voltage can be due to
   a. loss of residual magnetism.
   b. resistance greater than the critical field resistance.
   c. rotation of the armature in the direction opposite to that
      known to cause a voltage buildup.
   d. brush contact resistance effectively increasing the field
      circuit resistance above the critical point.
   e. improper connection of the field circuit leads at the brushes.
   f. all of the above

7. Voltage control refers to a change that takes place   A
   a. due to the operation of auxiliary regulating equipment.
   b. when the terminal voltage is increased.
   c. when the speed is regulated.
   d. automatically when the load is changed.

8. Voltage regulation refers to a change that takes place   C
   a. when speed is regulated.
   b. when the terminal voltage is increased.
   c. automatically when the load is changed.
   d. when auxiliary equipment is used.

9. When the load is raised from minimum to maximum, there is   _____C_____
    a. no change in terminal voltage.
    b. an increase in terminal voltage.
    c. a decrease in terminal voltage.
    d. less change than in other generators.

10. Connect the self-excited generator in Figure 3-2 by drawing the proper connections in the terminal boxes.

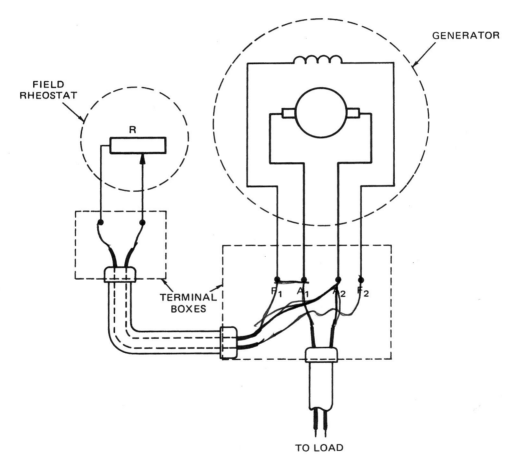

Figure 3-2 Connection diagram for self-excited shunt generator.

11. Write the formula for percent voltage regulation.

$$\frac{E_{nL} - E_{FL}}{E_{FL}} \times 100 = \text{\% regulation}$$

4

THE COMPOUND-WOUND DC GENERATOR

OBJECTIVES

After studying this unit, the student should be able to

- state the differences between a shunt generator and a compound-wound generator.
- define what is meant by a cumulative compound-wound generator and a differential compound-wound generator.
- describe how the voltage regulation of a generator is improved by compound windings.
- list changes in output voltage at full load due to the effects of overcompounding, flat compounding, undercompounding, and differential compounding.
- draw the basic generator circuit.
- connect the generator.

The voltage regulation of a generator is an important factor in deciding the type of load to which the generator should be connected. For lighting loads, a constant terminal voltage should be maintained when the load current increases. A simple shunt generator can only do this if expensive regulating equipment is also used.

Generators designed to maintain a constant voltage within reasonable load limits may have a double winding in the field circuit (Figures 4-1 and 4-2). The second winding is wound on top of, or adjacent to, the main winding. This second winding is called the *series winding* to distinguish it from the main shunt winding. The series winding has fewer turns than the shunt winding. Because the series winding is connected in series with the armature and load, it carries the full-load current; therefore, it is heavier gauge wire than the shunt field. A generator with such a *double-field winding* is called a *compound-wound generator*.

Figures 4-1 and 4-2 show the basic circuits of two ways to connect a compound-wound generator: the long shunt and the short shunt. In the short shunt circuit (Figure 4-1), the main shunt field is connected directly across the brushes; in the long shunt circuit (Figure 4-2), the shunt field is connected across the combination of the armature and the series field. The operating characteristics of these circuits are similar, but the short shunt is preferred because the shunt field remains more constant and is not affected by changes caused by the series field current.

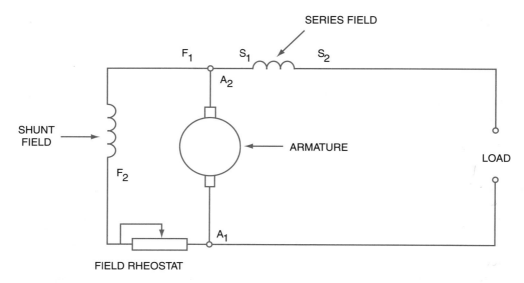

Figure 4-1 Short-shunt, compound generator connection.

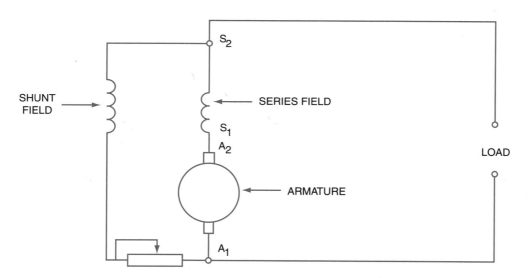

Figure 4-2 Long-shunt, compound generator connection.

COMPOUND FIELD WINDINGS

Two important details of the compound-wound generator must be considered: (1) the relative direction of the currents through both windings of a particular field pole, and (2) the magnetic effects that these currents can produce.

The series and shunt windings of a single pole of a compound-wound generator are shown in Figure 4-3. Winding (A) is the series winding through which the load current passes; winding (B) is the normal shunt winding. If the load current is in the direction shown in Figure 4-3, the magnetizing force of the series winding (A) will aid the shunt winding (B) and increase the strength of the magnetic field because the current in the shunt winding is not normally strong enough to saturate the core. If the load current through the series winding is in the direction opposite to that shown in Figure 4-3, its effect will be to weaken the magnetic field.

When the series winding is connected to aid the shunt winding, the generator is called a *cumulative compound-wound generator;* if the series winding is connected to oppose the magnetic shunt field, it is called a *differential compound-wound generator.*

The action of two fields in changing the flux density can be used to improve the voltage regulation of a normal shunt generator. As stated in Unit 1, as a load is applied in the shunt generator, the output voltage falls because of internal resistance, armature reaction, and the reduction of voltage applied to the field circuit. If the field strength can

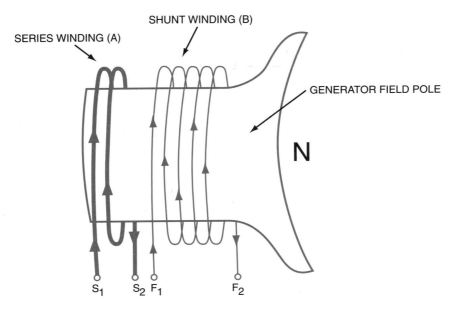

Figure 4-3 Compound field windings.

be *automatically increased in proportion* to load current as it increases, the output voltage can be maintained at a constant level. This is the objective in adding the series winding to the compound generator. As the load current increases in a cumulative-compound–connected generator, the load current passes through the series winding and increases the field flux. The additional voltage induced by cutting this flux compensates for the voltage loss due to armature resistance, armature reaction, and lower shunt field voltage. By adjusting how much current flows through the series winding, the output voltage can be more closely controlled. In fact, the output voltage could rise due to the increase in load current passing through the series field, thereby aiding the shunt field flux.

The number of turns in the series field helps determine the degree of compounding achieved. A large number of turns in the series winding produces *overcompounding* (a voltage increase at full load as compared to the output voltage at no load). A small number of series turns produces a reduced voltage at full load. This effect is called *undercompounding.*

Flat compound generators have the same voltage output at no load and full load. In industry, this type of generator is used where the distance between the generator and the load is short and line resistance is minimal. Overcompounding generators are used when the transmission distance is long, as in traction service, and the voltage at the end of the line must remain fairly constant.

A comparison of the voltage regulation of a shunt generator and a compound generator for both cumulative and differential connections is shown in Figure 4-4.

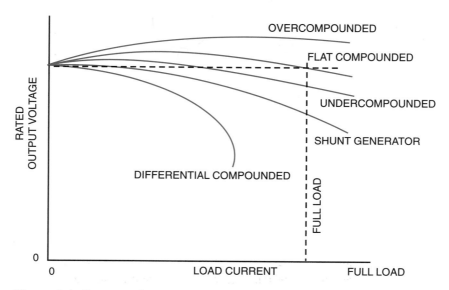

Figure 4-4 Compound generator load characteristics.

OUTPUT VOLTAGE CONTROL

The rated voltage of a compound generator operating at rated speed is set by adjusting the field rheostat. Because the compounding effect of the series field changes with speed, it is important to operate a compound generator at its rated speed.

Variation of Compounding

In general, compound-wound generators are designed by the manufacturer to have an overcompounding effect. The amount of compounding can be changed to any desired value by using a diverter rheostat across the series field. In Figure 4-5, a *diverter rheostat* (R) is connected in shunt (parallel) with the series winding. If the resistance of the diverter is set at a high value, the load current passes through the series winding to produce a maximum compounding effect. If the diverter is set at its minimum value, no load current passes through the series winding and the generator acts like a normal shunt generator. By adjusting the rheostat to intermediate values, any degree of compounding within these limits can be obtained. Flat compounding results when "no-load" voltage is equal to "full-load" voltage.

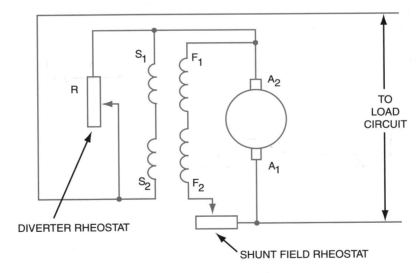

Figure 4-5 Diverter circuit.

SUMMARY

Compound-wound generators use a series field connected in series with the load to react to current load changes. It is connected to aid the shunt field or to oppose the shunt field. In most DC generators, the series field will be connected so that the magnetic field produced aids the shunt field flux. This connection is called a *cumulative connection*.

The degree of compounding can be controlled by a diverter rheostat. If the series field is connected so that the resultant flux opposes the shunt field flux, then the output voltage drops dramatically with an increase in current draw and the generator is differentially connected.

ACHIEVEMENT REVIEW

A. Select the correct answer for each of the following statements and place the corresponding letter in the space provided.

1. A compound-wound generator terminal connection box   _D_
    contains terminal leads
    a. F_1, F_2 and A_1, A_2.
    b. S_1, S_2 and F_1, F_2.
    c. S_1, S_2 and A_1, A_2.
    d. S_1, S_2, F_1, F_2, and A_1, A_2.

2. The series winding must be large enough to carry   _C_
    a. the total magnetic flux.
    b. a 300 percent overload.
    c. full line current.
    d. full line voltage.

3. Select the type of generator that may be used for loads   _A_
    quite distant from the generator.
    a. overcompounded
    b. flat compounded
    c. undercompounded
    d. differential compounded

4. The normal voltage of a compound generator is changed   _C_
    by adjusting the
    a. series field shunt.
    b. brush setting.
    c. shunt field rheostat.
    d. equalizer.

5. The resistance of a series field diverter should be   _C_
    a. comparatively high.
    b. equal to the resistance of the series field.
    c. a variable resistor.
    d. comparatively low.

6. To achieve a maximum compounding effect, the diverter   *B*
    rheostat should be
    a. set at its minimum value.
    b. set at a high value.
    c. set at a value midway between its minimum and
       maximum values.
    d. removed from the series field circuit.

B. Select the correct answer to questions 7 through 12 from the following list and
    place it in the space provided.

    a. field poles   g. increase
    b. diverter rheostat   h. flat compounding
    c. compound-wound generator   i. shunt field rheostat
    d. saturate   j. overcompounding
    e. decrease   k. remain constant
    f. shunt generator   1. undercompounding

7. When it is necessary to provide automatic control of the voltage output at constant
    speed, the generator selected is a *Compound wound generator*

8. The current through the shunt winding of a compound generator is not sufficient to
    _____*saturate*_____ the field poles.

9. The terminal voltage output of a cumulative compound-wound generator should
    _____*remain constant*_____ as the load current is increased.

10. When the output voltage of a generator is the same at both no load and full load,
    the generator is called a *Flat compounded* type.

11. Compound-wound generators are generally designed to be of the *overcompounding*
    type.

12. The amount of compounding that can be obtained from a generator is controlled
    by the _____*diverter rheostat*_____.

5

SUMMARY REVIEW OF UNITS 1–4

OBJECTIVE

- To provide the student with an opportunity to evaluate the information and understanding acquired in the study of the previous four units.

Select the correct answer for each of the following statements and place the corresponding letter in the space provided.

1. A generator series-field–diverter rheostat is always connected   _A_
   a. in parallel with the series field.
   b. in series with the series field.
   c. directly across the line.
   d. between the armature and the series field.

2. The voltage of an overcompounded generator   _D_
   a. decreases as the load increases.
   b. decreases as the load remains constant.
   c. increases as the load remains constant.
   d. increases as the load increases.

3. Brushes of a DC generator ride on the   _A_
   a. commutator.   c. shaft.
   b. armature.   d. commutating pole.

4. The field pole of a DC generator or motor is constructed   _B_
   with an iron core to
   a. decrease the magnetic flux.
   b. increase and concentrate the magnetic flux.
   c. decrease the residual magnetism.
   d. increase the eddy currents.

5. A generator with terminal markings A_1–A_2, F_1–F_2, S_1–S_2 is a   _D_
   a. separately excited generator.   c. series generator.
   b. shunt generator.   d. compound generator.

6. The field core of a DC generator is   _B_
   a. the round part of the rotating field.
   b. wound with wire.
   c. usually round on large machines.
   d. the part of the generator that holds the armature in place.

7. In self-excited DC generators, initial field excitation   _D_
   is produced by
   a. current in the coils.   c. magnetic flux.
   b. moving the field.   d. residual magnetism.

8. DC generators and motors have   _C_
   a. one pole.   c. pairs of poles.
   b. two poles.   d. four poles.

9. To reestablish the magnetic field of a generator where   _C_
   the magnetic field has been lost, you must
   a. run the generator above rated speed.
   b. reverse the connections to the shunt field.
   c. apply DC to the shunt field coils.
   d. attach a permanent magnet to the armature.

10. If a generator is overcompounded and you want to decrease the degree of com-
    pounding, then the diverter resistance must be ___~~increased~~___ . (increased,
    decreased)   _decreased_

11. Write the formula for calculating the percent voltage regulation in a DC generator.

    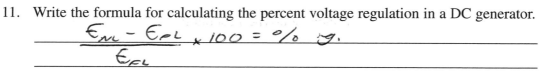

    $$\frac{E_{NL} - E_{FL}}{E_{FL}} \times 100 = \% \text{ reg.}$$

12. The preferred connection for DC compound generators is the _short-shunt_
    connection. (short-shunt, long-shunt)

13. The series field winding is a _larger_ size wire than the shunt field winding.
    (larger, smaller)

6

SINGLE-PHASE AC GENERATION PRINCIPLES

OBJECTIVES

After studying this unit, the student should be able to

- explain how AC voltage is generated.
- state the differences between a stationary armature and a stationary field generator.
- determine how to control the amount of output voltage.
- calculate the output frequency of an AC generator.
- explain what is meant by a single-phase sinewave.

BASIC PRINCIPLES OF AC GENERATION

Some basic principles are used in the generation of AC (alternating current) voltage. The major source of electrical power is provided by electromagnetic generation. This method uses the principle of moving a magnetic field past an electrical conductor. This induces a voltage in the conductor and creates a resultant current flow to an electrical load. For electromagnetic induction-type generators, the following three factors are needed for induction:

- The presence of a magnetic field
- An available electrical conductor
- Relative motion between the conductor and the magnetic field

In Figure 6-1, the magnetic field is moved past the stationary conductor. As the magnetic field moves past the conductor, the magnetic lines of force, called *flux,* pass through the conductor or *cut* the conductor. The equation associated with Faraday's law states that if 10 million lines of magnetic flux cut through a conductor in 1 second, then 1 volt is induced into the conductor. Assuming the electron flow theory is used (where electron current flow moves from negative to positive), then the left-hand rule for generators is used, as shown in Figure 6-2. The resultant current flow to the electrical load is toward you. If the direction of motion is reversed, the resultant direction of electron

Figure 6-1 Direction of induced EMF and current flow as a conductor is moved through a magnetic field.

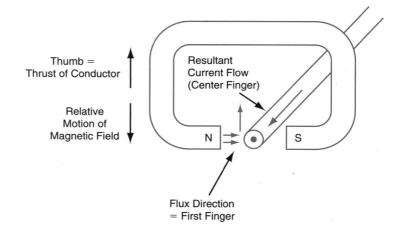

Figure 6-2 Left-hand rule for generators used with electron flow theory.

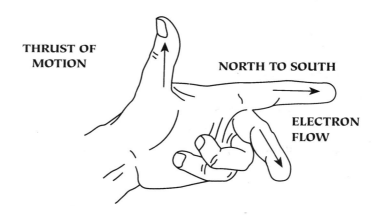

flow is reversed. Likewise, if the magnetic field is reversed and the direction of motion is constant, then the resultant electron flow is reversed.

Stationary Field

A more practical explanation of electron flow can be shown with the use of an electrical generator. If a conductor is spinning inside a generator (with a magnetic field established on the stationary portion of the generator), then the conductor will pass through stronger and weaker magnetic fields and in different directions. As the conductor passes directly under the center of the magnetic field pole, the flux is the most dense and the induced voltage is greatest. This condition creates the most induced EMF (electromotive force) and produces the peak of the AC waveform. As the conductor moves parallel to the lines of flux, no *cutting* action takes place and no voltage is induced. This represents the zero voltage point of the AC waveform. If you multiply the sine of the

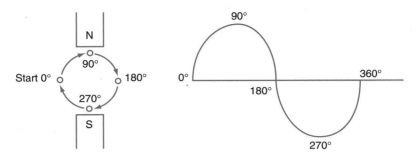

Figure 6-3 Mechanical generation of voltage and the resultant sinewave.

angle that the conductor moves from the horizontal starting point (zero degrees) times the maximum induced voltage in the conductor, you can plot a graph of each induced voltage level at each angular position. The graph is a sinewave representation of the induced voltage representing the induced AC voltage (see Figure 6-3).

Frequency

This example moves one conductor 360 mechanical degrees to produce 360 degrees of sinewave or one complete cycle. If the conductor spins 60 times in 1 second, 60 complete cycles of the waveform are produced. This is called 60 cycles per second or 60 hertz (Hz). Revolution of the conductor is 60 revolutions per second, or 3,600 revolutions per minute.

Stationary Armature

Another method is most often used to generate a sinewave. This method spins a magnetic field inside a looped conductor. The looped conductor will multiply the effects of the magnetic field cutting just one conductor. Figure 6-4 illustrates a method of generation referred to as a stationary-armature, rotating-field type generator. In this case, the magnetic field moves past the conductors, *cutting* through them and inducing a voltage into the stationary conductors. The waveform on the conductors is still the same AC sinewave because the magnetic field is moving past the conductors with varied strength in relative positions.

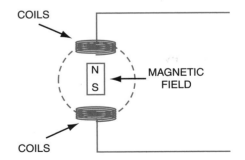

Figure 6-4 Mechanical generation using a stationary armature and a rotating magnetic field.

When a generator is used to produce a single sinewave output, it is referred to as a single-phase generator. It is only designed to produce a single sinewave of AC output voltage.

Frequency is a term used to describe how often or how frequently a complete AC waveform is produced. The frequency of a generator that produces 60 complete sinewave cycles in 1 second is 60 Hz. The formula used to determine the frequency of the generated waveform is

$$\text{Frequency} = \frac{\text{RPM} \times \text{number of poles}}{120}$$

Frequency is measured in hertz, or cycles per second.

RPM is the speed of the rotating member, either the conductor or the magnetic field.

Number of poles is the number of magnetic poles of the magnetic field (in this example, 2 poles).

The number 120 is a constant used to convert RPM to revolutions per second, and pairs of magnetic poles to number of poles used in the formula.

Single-Phase Generator Maintenance

Maintaining single-phase generators is relatively simple from the electrical standpoint. Most of the problems that occur are from the prime mover, or the engine input, that drives the generator. The maintenance of the generator itself depends on the actual style of generator. The generator is either a stationary field or a stationary armature. The points to remember on the generation side are (1) a magnetic field must be present; (2) motion between the field and the conductors of the armature must exist; and (3) the output electrical power must be connected to the load. Therefore, these three factors should be checked if the generator does not have correct output. (See more details on three-phase generators in Unit 11.)

Single-Phase and Three-Phase Generation

Single-phase power systems are used extensively, but large power generation systems typically generate three phases at a time. This is used to supply three distinct sinewaves to a power system. (See Unit 7.) The three waveforms or three phases can be used to supply three-phase motors, or the power needs can be split up to provide three separate single-phase systems.

SUMMARY

The main method of generation is through electromagnetic induction. Either a magnetic field is moved past the stationary conductor, or the conductor is moved through the magnetic field. As the density of the magnetic flux is greater at the center of the magnetic field, there is more induced EMF and a peak value of voltage is produced. As the relative direction of motion between the conductor and the magnetic field changes, the direction of the induced EMF changes. The number of complete sinewaves produced in a generator per second is referred to as the frequency and is measured in hertz (Hz).

ACHIEVEMENT REVIEW

1. Calculate the frequency of a single-phase generator that has four poles and spins at 1,800 RPM.

$$F = \frac{RPM \times poles}{120}$$

$$F = \frac{1800 \times 4}{120} = 60\ hz$$

2. What is the unit of measure of frequency, and what does it refer to? _____

   Frequency is the number of spins
   a conductor makes in a second
   Frequency is measured in hertz

3. Why is the standard AC waveform referred to as a sinewave? _____

   when the degree of motion of a conductor
   is relative to the amount of
   flux it creates is plotted on a graph it looks
   like a sinewave

4. Explain what is meant by using the left-hand rule. _____

   when one finger points in the direction
   of flux, the thumb points in the direction
   of motion, so the other finger points
   in the direction of current flow

U • N • I • T

7

INTRODUCTION TO POLYPHASE CIRCUITS

OBJECTIVES

After studying this unit, the student should be able to

- define what is meant by polyphase systems.
- state the advantages in the generation and transmission of three-phase power.
- measure and calculate power in three-phase systems.
- calculate the power factor in three-phase systems.

Almost all power transmission uses the three-phase system. In the three-phase system, electrical energy originates from an alternator that has three main windings placed 120° apart. A minimum of three wires is used to transmit the energy generated.

A polyphase system, therefore, is the proper combination of two or more single-phase systems. In their order of usage, the most common types of polyphase systems follow:

- three phase (used for power transmission)
- six phase (used for power rectification)
- two phase (used for power rectification)

Figure 7-1 shows equipment assembled in a transmission line substation, and Figure 7-2 shows a three-phase generator in a utility power-generating station.

ADVANTAGES OF THREE-PHASE SYSTEMS

The advantages of three-phase systems apply to both the generation and transmission of electrical energy.

Generation

A three-phase generator can be compared to a gasoline engine. An eight-cylinder engine develops eight small pulses of power per cycle as compared to one large surge of power per cycle for a one-cylinder engine. Similarly, a three-phase generator induces voltage in three windings per turn, rather than in just the one winding of a single-phase

Figure 7-1 A transmission line substation.

Figure 7-2 A steam turbine electric generator used at a utility power-generating station.

generator. In addition, the generator actually is smaller in physical dimensions than a single-phase generator of the same rating. Three-phase generation produces energy more smoothly than single-phase generation and provides for more economical use of space within the frame of the machine.

GENERATION OF THREE-PHASE ELECTRICAL ENERGY

Figure 7-3 shows the arrangement of the windings in a simple three-phase AC generator. The coils are spaced 120 electrical degrees apart. The voltage diagram shows the relationship of the instantaneous voltages as the rotating field poles turn in the direction indicated.

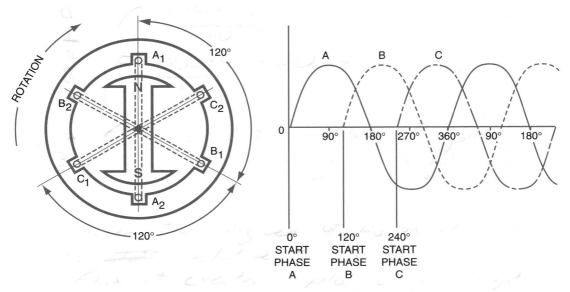

Figure 7-3 Generation of three-phase electrical energy.

Three-Phase Winding Connections

The internal winding connections shown in Figure 7-4 for a three-phase generator are arranged so that any of three or four wires may be brought out. That is, three-phase windings may be connected either in the star (wye) pattern (Figure 7-5), or the delta pattern (Figure 7-6). (See Units 8 and 9.)

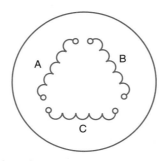

Figure 7-4 Schematic diagram of three-phase windings.

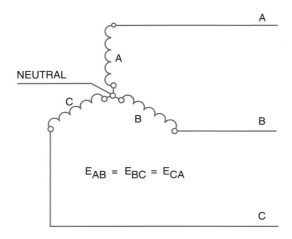

$$E_{AB} = E_{BC} = E_{CA}$$

Figure 7-5 Wye three-phase connection.

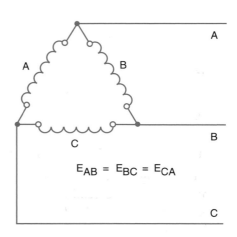

$$E_{AB} = E_{BC} = E_{CA}$$

Figure 7-6 Delta three-phase connection.

SIX-PHASE CONNECTIONS

Six-phase power is usually applied to power rectifiers. The AC six-phase supply is converted from a three-phase power line by a bank of three transformers connected for six-phase on the secondary side of the line transformer. The double-wye, six-phase connection shown in Figure 7-7 is one method of obtaining six-phase power. The lines are brought out at the outer ends of the windings. The points (1-2-3-4-5-6) are displaced 60 electrical degrees apart from one another.

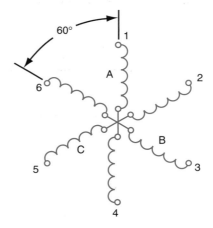

Figure 7-7 Six-phase connection (double wye).

TWO-PHASE CONNECTIONS

In a two-phase connection (Figure 7-8), the windings are spaced 90° apart. Lines A_2 and B_1 are often connected to form a three-wire, two-phase system. The A and B phase voltages are designed to be equal.

Transmission Using Transformers

Three-phase transmission saves material, installation time, and maintenance costs. A three-phase, 4-wire WYE system can provide

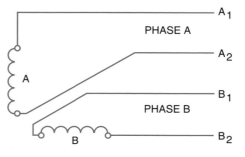

Figure 7-8 Two-phase connection.

three 120-V lighting circuit lines, three 208-V single-phase circuits, and one 208-V, three-phase power line over four wires.

MEASUREMENT OF THREE-PHASE POWER

Three-phase power may be measured using either the two-wattmeter method or a polyphase wattmeter.

Two-Wattmeter Method (Using Blondell's Theorem)

The reversing switches of both wattmeters must be set in the same direction. If one meter shows a negative reading, reverse the corresponding switch so that a positive reading is seen on the scale. If the switches are set in opposite directions, the lower reading is assumed to be negative.

If the power factor (PF) is 100 percent, $W_1 = W_2$, then $W_T = W_1 + W_2$.

If the PF is greater than 50 percent and less than 100 percent, then W_1 and W_2 are unequal. $W_T = W_1 + W_2$. (W_T = Total power.)

If the PF is 50 percent, one meter reads 0 and $W_T = W - 0$.

If the PF is less than 50 percent, one meter has a negative reading and $W_T = W_1 - W_2$.

Note: The two-wattmeter method cannot be used in an unbalanced three-phase, four-wire system. See Figure 7-9 for the two-wattmeter connection.

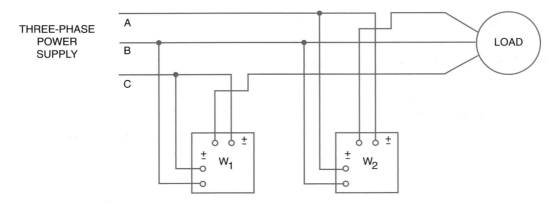

Figure 7-9 Measurement of three-phase power using the two wattmeter method.

Polyphase Wattmeter Method

The connections of a polyphase wattmeter used to measure three-phase power are shown in Figure 7-10. In the polyphase wattmeter, the torque produced by two current coils and two voltage coils causes the pointer to deflect and indicate total watts in the circuit in one instrument.

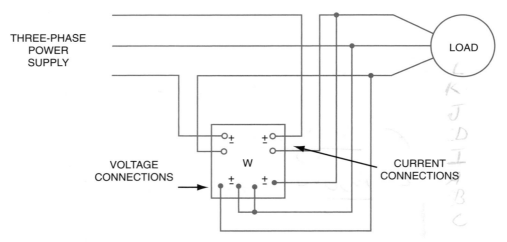

Figure 7-10 Polyphase wattmeter connections for measurement of power in a three-phase circuit.

CALCULATION OF THREE-PHASE POWER AND THE POWER FACTOR

Power Factor

The power factor is the ratio of true power to apparent power. A power factor of 100 percent is the best electrical system.

$$PF = \frac{\text{True Power}}{\text{Apparent Power}} = \frac{\text{3-phase watts}}{\sqrt{3} \times E_L \times I_L} = \frac{\text{watts}}{\text{voltamps}} \quad \text{Note:} \sqrt{3} = 1.73$$

This formula can be used only for *balanced three-phase circuits*.

A power factor meter can be used to measure the power factor in a three-phase circuit for both balanced and unbalanced conditions.

Power, then, is calculated using the following expression:

$$\text{3-phase watts} = \sqrt{3} \times E_{\text{Line}} \times I_{\text{Line}} \times PF$$

The substitution of values in this formula determines the true power in a three-phase *balanced* circuit only.

Three-phase equipment is designed to operate as a balanced load. A three-phase circuit containing a combination of single- and three-phase loads is very seldom balanced. A polyphase wattmeter *must* be used for the measurement of power in an unbalanced circuit. See Figure 7-11 for an example of a three-phase power analyzer that measures watts, VARs, VA, and so forth.

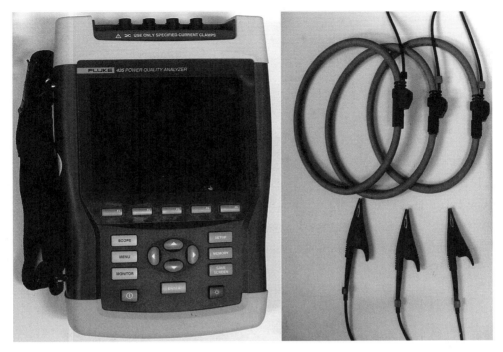

Figure 7-11 A three-phase power analyzer has voltage leads and current sensors to measure watts, VARs, VA, and other circuit conditions.

SUMMARY

Polyphase refers to any electrical system that has more than one waveform present. Typically, three sinewaves are produced by commercial generation equipment. These three sinewaves are separated by 120 electrical degrees. The generator coils may be connected either in the wye pattern or the delta pattern. Other connections such as the six-phase and the two-phase are also possible. Measurement of three-phase power can be accomplished by using a two-wattmeter method and the appropriate formulas, or by using the correctly connected polyphase wattmeter.

ACHIEVEMENT REVIEW

For each of these numbered items, select the letter of the phrase from the following list that will complete the statement. Write the letter in the space provided.

a. two wattmeters
b. 3-phase watts = $\sqrt{3} \times E \times I \times PF$
c. polyphase wattmeter
d. three wires
e. four wires

f. five wires
g. six wires
h. more economical
i. rectification
j. two-phase power

k. three-phase power   n. lighting
1. more than one phase   o. E × I × PF
m. 180°

1. Polyphase means   _L_
2. Three generator windings spaced 120° apart generate   _K_
3. Two generator windings spaced 90° apart generate   _J_
4. Three-phase power is transmitted over a minimum of   _D_
5. The six-phase connection is used for   _I_
6. Three-phase power may always be measured by using a   _A_
7. The formula for calculation of three-phase power is   _B_
8. Unbalanced three-phase power can be measured with a   _C_
9. The three-phase system is better than the single-phase line
   because it is   _H_
10. Lighting circuits usually are connected to a three-phase line having   _E_

8

THE THREE-PHASE WYE CONNECTION

OBJECTIVES

After studying this unit, the student should be able to

- diagram the proper connections for a wye-connected generator and transformer.
- state the applications of the wye-connected generators and transformers in three-phase distribution systems.
- compute the voltage and current values in various parts of the wye-connection circuit.

The star or wye connection is particularly suited for the distribution of power and lighting where one three-phase transmission line supplies the energy. All three transformers in the wye bank can supply single-phase loads as well as a three-phase load. The wye system also provides a grounded neutral with equal voltage between each phase wire and the neutral.

The two standard voltage patterns available from the four-wire wye system in the United States are 120/208 and 277/480. The 120/208-volt wye system is most commonly used for small industrial plants, office buildings, stores, and schools. In these applications, the main electrical need is for 120-volt lighting and equipment circuits, and only a moderate amount of 208-volt three-phase power load. The 277/480-volt wye system is mainly used for large commercial buildings and industrial plants where there is a higher demand for power at 480 volts, three phase, and lighting at 277 volts, single phase.

The types of three-phase systems are named for the diagram shape of the transformer secondary winding connections. The wye or star system, as shown in Figure 8-1,

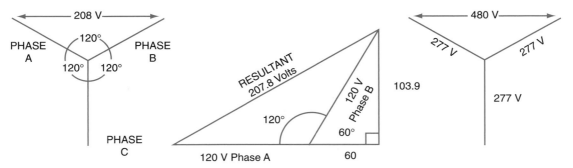

Figure 8-1 Vector diagram shows resultant voltage when adding coil voltages in wye system.

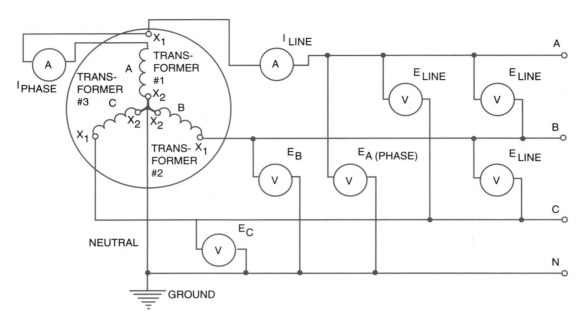

Figure 8-2 Wye connection in an alternator, or three-phase transformer bank.

is shaped like the letter Y. The wye or star connections are made by tying together the ends of the three transformer windings, labeled X_2, and bringing this termination out as the *neutral* wire. The remaining three unidentified conductors of the four-wire, three-phase system, labeled on the figure as A, B, C, are tied to the three X_1 ends, respectively. Alternators are connected in the same manner as shown in Figure 8-2.

VOLTAGE RELATIONS

The voltage reading across any pair of line wires of a balanced three-phase wye connection is equal to the vector sum of the two-phase windings connected in series across the pair of lines.

For example, if the phase winding voltage is 120 volts, the line voltage is 208 volts.

$$E_{Line} = 1.73 \times 120 = 208 \text{ V}$$

The voltage from any line to a grounded neutral is the phase winding voltage and is usually called the *phase voltage*. Phase is represented by the Greek letter phi (ϕ).

Figure 8-1 illustrates the relationship between the voltage in a wye or star system. The line-to-line voltage is the vector sum of the individual coil voltages. The coil voltage is the voltage generated by an individual phase coil, either in a generator or a coil winding in a transformer. By connecting them into a wye pattern, the voltages add vectorially. Because the sinewaves are 120° apart in each phase, the two phases are not added arithmetically. Figure 8-1 shows the vector addition of two 120-volt phase voltages added together to get a resultant line voltage of 207.8 or a nominal 208 volts line to line. The same relation exists in other wye connections. Phase voltage multiplied by the constant 1.73 yields the line voltage. In a 277/480-volt wye system, the same relation exists. The phase voltage, 277 volts, multiplied by 1.73 yields the line voltage of 480 volts nominal.

Another common factor is used when working with three-phase voltages. When converting from line voltage to phase voltage, divide the line voltage by 1.73 to get the phase voltage. For example, 480 volts line to line divided by 1.73 yields 277 volts nominal. Another common factor is the inverse of 1.73 (.58), which allows the electrician to multiply the line-to-line voltage by .58 to obtain phase voltage. It is an easier factor to use for approximating the phase values in a quick calculation.

CURRENT RELATIONS

The line current is the same as the phase current in a wye connection because each phase winding is connected in series with its corresponding line wire (Figure 8-2). Remember that current in a series circuit is the same throughout all parts of the circuit.

$$I_{Line} (I_L) = I_{Phase} (I_\phi)$$

To calculate the current in the common conductor of a three-phase four-wire wye system, we must examine how the currents combine from all three-phase wires. To analyze a balanced condition, consider that the common conductor will carry the imbalance of the three-phase, or hot, conductors. If the three phases are equal, as in supplying a three-phase motor, the neutral current is the imbalance of the three-phase conductors. In this case there is no imbalance and the neutral current is zero. The formula used to find the neutral current in a three-phase wye system is

$$I \text{ neutral} = \sqrt{A^2 + B^2 + C^2 - AB - AC - BC}$$

You can quickly tell what the neutral current is if all the phase currents are equal. For instance, if they are all 10 A, then

$$I \text{ neutral} = \sqrt{10^2 + 10^2 + 10^2 - 100 - 100 - 100} = 0 \text{ amps}$$

To find the neutral current in a three-phase system when the phases are not equal, we use the same formula and perform a calculation. If the phase currents are different, such as A phase is 5A, B phase is 10 A, and C phase is 12A, then the formula is

$$I \text{ neutral} = \sqrt{5^2 + 10^2 + 12^2 - 50 - 60 - 120} = 6.25 \text{ A}$$

Another combination may exist when dealing with three-phase power and single-phase devices. If you use two of the three phases of a four-wire wye and the neutral to operate a single-phase load, then the same formula is used to find the actual current in the neutral. This situation is *not* the same as a single-phase power source with two hots and a neutral conductor. In the single-phase situation, the neutral carries only the imbalance of the two hots. In the three-phase wye situation, the neutral formula must be used. For instance, if two hot conductors of a three-phase system each provide 10 A, then the neutral formula is

$$\text{I neutral} = \sqrt{10^2 + 10^2 + 0^2 - 100 - 0 - 0} = 10 \text{ A}$$

The neutral carries the same as the hot conductors. If the hots carry different values, the formula must be used to calculate the neutral current.

POWER RELATIONS

The power formula used for three-phase, wye-connected systems is the same as that used for the delta-connected system. The relationship of the current and voltage is taken into account. The three-phase apparent power for a three-phase system can be calculated by finding the individual-phase apparent power and multiplying by 3 for the power of a balanced three-phase system. Phase current × phase voltage × 3 will yield the three-phase apparent power measured in volt-amps. When using the line voltage values to find three-phase power, the phase voltage has already been increased by a factor of 1.73 to get the line voltage. Another multiplying factor of 1.73 must be used to equal the same product found when using the three single phases in the power calculations. Therefore, the formula for three-phase apparent power is line voltage × line current × 1.73, which yields the same apparent power answer as previously calculated.

3-phase apparent power: line voltage × line current × 1.73 = 3-phase volt-amps

To calculate three-phase true power, which is measured in watts, the power factor of the AC system must be known. As discussed previously, the power factor is a factor multiplier to the apparent power of an AC system that yields the true power in watts. As a reminder, power factor is the cosine of the phase angle that is established between the voltage and the current in an AC system that is not totally resistive in nature. To find the true power of a three-phase power system, the additional power factor is used as follows:

3-phase true power: line voltage × line current × 1.73 × power factor = 3-phase watts

APPLICATION

Typical applications of three-phase, wye-connected systems use voltage ratings of 120/208 where the 120 designates the single-phase voltage, line to neutral, and the 208 volts designates the line-to-line voltage. The 208-volt, three-phase voltage is used in commercial applications for three-phase motors. Another common voltage in the United States is 277/480-volt systems. The 277-volt designation is the single-phase line to neutral and is used extensively for commercial lighting applications. The 480-volt value is the line-to-line voltage and is used extensively as three-phase commercial power to buildings.

A four-wire transmission line usually originates from a transformer bank or a generator connected in wye. Both lighting and power circuits are connected to the four-wire system. Four circuits are served by four wires.

Lines	A-B-C	Power
Lines	A-Neutral	Three lighting circuits
	B-Neutral	
	C-Neutral	

In summary, the following statements are true of the three-phase wye connection:

$$E_{Line} = 1.73 \times E_{Phase}$$
$$I_{Line} = I_{Phase}$$

Lines A-B-C supply power circuits.
Lines to neutral serve lighting circuits.

Either three- or four-wire transmission lines are attached to a wye-connected generator or transformer bank.

WYE CONNECTION OF TRANSFORMER WINDINGS

A bank of three transformers can be connected in wye, delta, or other three-, six-, twelve-, or eighteen-phase arrangements. Figure 8-3 shows a conventional method of connecting three transformers in a three-phase wye arrangement. Compare this with Figure 8-2. (Note that X_2 of each coil is connected to neutral.)

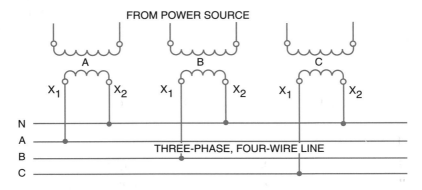

Figure 8-3 Transformer bank with a wye connection.

SUMMARY

One method of connecting three-phase coils is to connect them into a pattern that schematically resembles the letter Y. This pattern of connection is referred to as the *wye* or the *star* connection. In the wye connection, the waveforms are 120° apart and there are three distinct sinewaves. The resultant voltage, obtained by connecting these single phases together, is to increase the line-to-line voltage by a factor of 1.73. In this pattern the line voltage is 1.73 times the phase voltage, but the line current is the same as the phase current. Three-phase volt-amp capacity is calculated as follows:

$$\text{Line E} \times \text{Line I} \times 1.73$$

Figures 8-4 and 8-5 show transformers connected to power generating substations. The secondary of the transformers would be connected in a wye pattern.

Figure 8-4 Unit substation in individual enclosures.

Figure 8-5 Three-phase, high-voltage transformer bank.

ACHIEVEMENT REVIEW

1. Indicate the number of the leads that must be connected to make a wye connection in Figure 8-6.

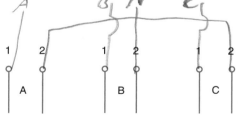

Figure 8-6 Lead connections for question 1.

4 leads   X1 X2 X3 X0 -neutral

2. Where will the neutral be brought out in the diagram (Figure 8-6) shown in question 1? Describe or use a sketch.

the neutral is brought out from the center tap - Xo

3. What two types of circuits are supplied by the three-phase, four-wire system?

    120/208 , 277/480

4. Make a complete wye connection in Figure 8-7.

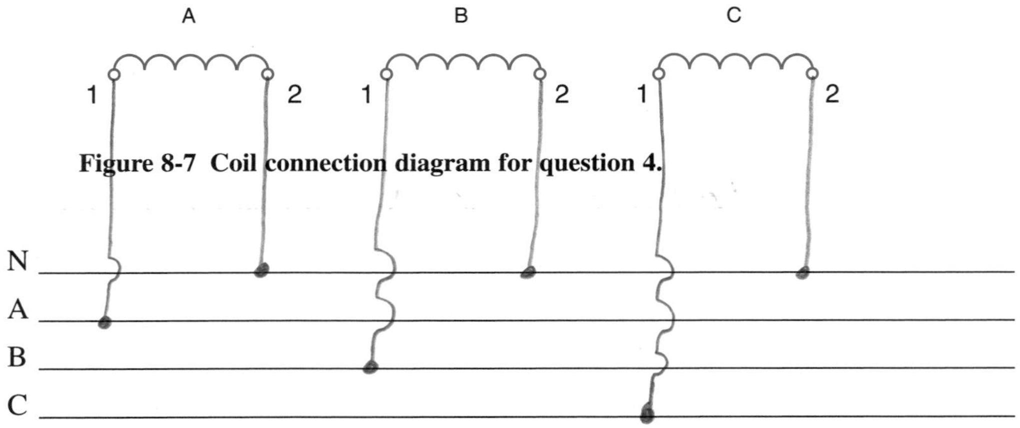

Figure 8-7 Coil connection diagram for question 4.

5. $E_{line} = 1.73 \times E_{phase}$
   The phase current and phase voltage of each winding of an AC generator are 10 amperes and 100 volts, respectively. Determine the line voltage and current.

   $100 \times 1.73 = line$ $E_{line} = 173 V$

   $I_{line} = I_{phase}$ $I_{line} = 10 A$

The following circuit is incomplete. Questions 6 through 10 are based on the circuit displayed in Figure 8-8.

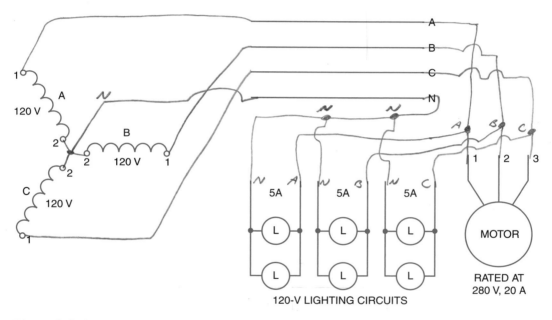

Figure 8-8 Connection diagram for questions 6 through 10.

6. Complete the connections for a four-wire wye system.

7. Determine the line voltage.

   $E_{line} = 1.73 \times E_{phase}$

   $E_{line} = 1.73 \times 120 = 208 V$

8. Connect the motor for three-phase operation.

9. Connect the lamp banks for a balanced three-phase load.

10. Determine the phase current when only the lighting load is on the line.

    $I_{phase} = I_{line}$ $I_{phase} = 5 A$

9

THE THREE-PHASE DELTA CONNECTION

OBJECTIVES

After studying this unit, the student should be able to

- diagram the proper way to make a delta connection.
- state the applications of a delta-connected circuit in three-phase distribution systems.
- compute the voltage and current values in various parts of the delta-connection circuit.
- make a delta connection.

The delta connection, like the wye connection, is used to connect alternators, motors, and transformers. Delta is a Greek letter, which is shaped like a triangle (Δ). The delta connection takes its name from this symbol because of its triangular appearance. The schematic diagram of the winding connection of an alternator or secondary transformer bank shows the windings that are actually spaced 120 electrical degrees apart (Figure 9-1).

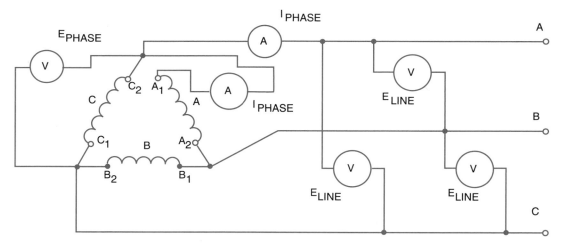

Figure 9-1 Delta connection in an alternator (or three single-phase transformers).

CONNECTIONS

To make a delta connection, connect the beginning of one phase to the end of the next phase until the last and final connection is to be closed. *Do not complete the delta connection* until the voltage is measured across the last two ends (see C_2–A_1 in Figure 9-2).

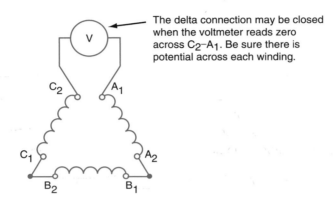

The delta connection may be closed when the voltmeter reads zero across C_2–A_1. Be sure there is potential across each winding.

Figure 9-2 Test for completion of the delta connection.

Test for Completion of the Delta Connection

If the voltmeter reads zero across C_2–A_1, the circuit may be closed (Figure 9-2). If the voltmeter reads twice the voltage of the phase winding, reverse any phase and retest. If a potential remains across C_2–A_1, reverse a second phase and make a final voltage test before completing the delta connection. The phase windings must have potentials 120 electrical degrees apart.

VOLTAGE RELATIONS

The voltage measured across any pair of line wires of a balanced three-phase delta connection is equal to the voltage measured across the phase winding (see Figure 9-1).

$$E_{Line} = E_{Phase}$$

CURRENT RELATIONS

Trace any line, such as line A in Figure 9-1, back to the connection point of phases C and A in a closed delta. The current in line A is supplied by phases A and C at the point of connection in the AC generator. Phases A and C are out of phase by 120°. The line current, therefore, is the vector addition of the two phase currents. In a balanced circuit, the phase currents are equal. The line current is determined by the following formula:

$$I_{Line} = \sqrt{3} \times I_{Phase} \text{ or } 1.73 \times I_{Phase}$$

For example, if the phase current in each winding of a generator or transformer is 10 amperes, the line current is equal to $1.73 \times 10 = 17.3$ amperes.

$$I_{Line} = 1.73 \times 10 = 17.3 \text{ amperes}$$

In a four-wire delta connection, the neutral currents are calculated differently than in a wye system. Because the neutral is positioned as a midpoint of a one-phase winding, the current in the neutral is calculated the same way as the current in a single-phase system where there are two hot conductors and the neutral carries the imbalanced current. For instance, if the two-phase conductors on either side of the neutral are equal currents of single-phase load values, the neutral current is zero. The formula is

$$|A \text{ phase} - C \text{ phase}|_{\text{single phase condition}} = \text{Neutral current}$$

Example If the A phase is 20 A and the C phase is 15 A, then the neutral is 5 amps. If using all three phases in the delta, subtract the balanced three-phase currents from the values for the A and C phases. The neutral connection current is still the imbalance between the A phase and the C phase.

Example If there is single-phase and three-phase load on a delta four-wire system and phase A has 30 A, phase B has 10 A, and phase C has 16 A, we could assume the three-phase load is 10 A. Then subtract the 10-amps of three-phase current from all phase values to yield A − 20 A, B − 0 A, and C − 6 A. The formula for the neutral current is

1256 − 300 − 160 − 480
436 − 0 − 0 − 120

$$: |A \text{ phase} - C \text{ phase}|_{\text{single phase condition}} = \text{Neutral current}$$

Therefore, $|20 - 6|_{\text{single phase condition}} = 14$ A of neutral current.

APPLICATION

The delta connection may be used as the source of a three-wire transmission line or distribution system. The three-wire delta system is used when three-phase power on three conductors is required.

In summary, the following statements are true of the three-phase delta connection:

$$E_{Line} = E_{Phase}$$
$$I_{Line} = 1.73 \times I_{Phase}$$

DELTA CONNECTION OF TRANSFORMERS

Figure 9-3 shows the delta secondary connection of a bank of transformers.

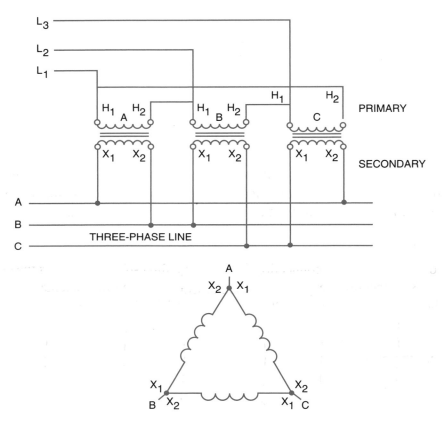

Figure 9-3 Delta connection for a transformer bank.

FOUR-WIRE DELTA CONNECTION

A transformer secondary can also be connected as a four-wire delta secondary. In this pattern (see Figure 9-4), one of the transformer secondary windings has a connection point brought out, creating a fourth wire connection. This point is at the midpoint of the winding and is referred to as the *center tap*. This yields a voltage that is half of the single-phase voltage. The connection point to the center tap is normally grounded. This connection is typically on the B phase, so the voltage from line leads A or C to the center tap is half the full transformer secondary. The voltage to the line wire B is actually 1.73 × the center-tapped voltage. As a typical application, the transformer secondary might be a 240-volt three-phase delta connection, which means 240-volt phase to phase. From line wire A to B to C, the voltage you would measure would be 240 volts. From A

Figure 9-4 Connection pattern for four-wire delta transformer.

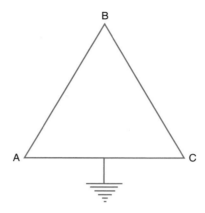

or C to the grounded center tap would measure 120 volts, and the voltage from B to the grounded center tap would be 208 volts. This is officially known as "the phase with the higher voltage to ground." See Unit 20 for connections and applications.

SUMMARY

A common method of connection for coils of a three-phase system is the delta connection. This connection pattern gets its name from the way the schematic representation of the connection represents the Greek letter delta (Δ). In this pattern, the three-phase coils are connected so that the line voltage is the same as the coil or phase voltage. However, the line current is the vector sum of the coil currents and is determined by multiplying the phase current by 1.73. The three-phase volt-amp capacity of the delta-connected, three-phase system is calculated as follows:

$$\text{Line E} \times \text{Line I} \times 1.73$$

ACHIEVEMENT REVIEW

1. As shown in Figure 9-5, six leads are brought out of a three-phase alternator and marked as indicated. Connect these six leads to make a three-phase delta connection.

Figure 9-5 Connection diagram for question 1.

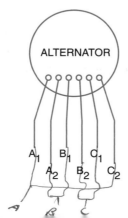

2. If the rated line voltage of an alternator is 120 volts, how can the alternator be connected for a rated voltage of 208 volts? _____

_____

_____

_____

3. A wye-connected alternator is rated at 20 amperes of line current. The internal connections are changed from wye to delta. What is the new line current rating?

_____

_____

_____

_____

4. Connect the three-phase windings in delta in Figure 9-6.

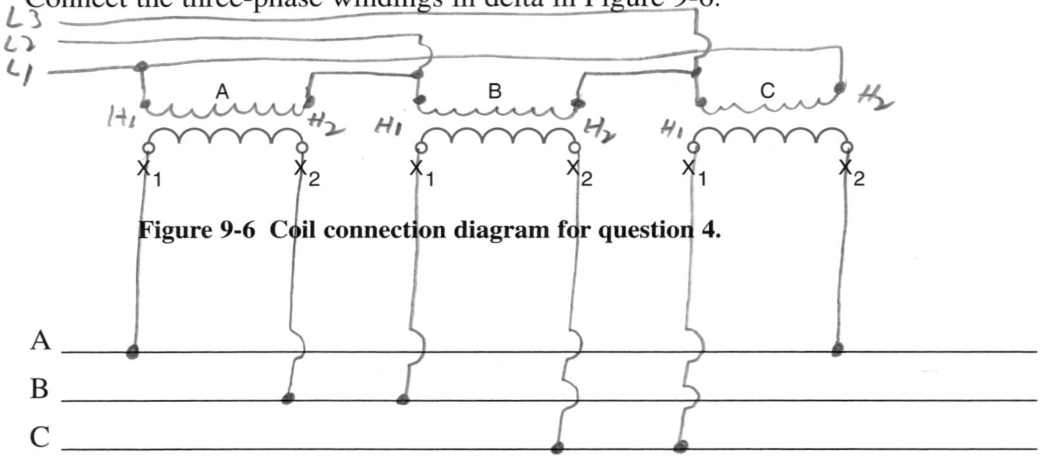

Figure 9-6 Coil connection diagram for question 4.

A _____

B _____

C _____

5. What precautions must be taken before closing a delta connection?

_____

_____

_____

_____

_____

_____

_____

Questions 6 through 10 are based on Figure 9-7.

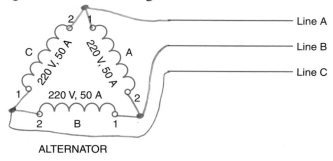

Figure 9-7 Coil pattern for questions 6 through 10.

6. Complete the connections to the line wires for a three-wire delta system.

7. Determine the rated line current of this three-phase line. $\underline{I_{Line} = 1.73 \times I_{phase}}$
    $\underline{I_{Line} = 1.73 \times 50 = 86.5A}$

8. What is the line voltage? $\underline{E_{line} = E_{phase} \quad E_{Line} = 220v}$

9. Why is this connection called a delta connection? _____
    It's shaped like a triangle or the greek
    letter delta (Δ).

10. Why is the delta connection limited to three-wire, three-phase transmission circuits?
    It's shaped like a tr

U•N•I•T
10

SUMMARY REVIEW
OF UNITS 6–9

OBJECTIVE

- To provide the student with an opportunity to evaluate the information and understanding acquired in the study of the previous four units.

Select the correct answers for each of the following statements and place the corresponding letter in the space provided.

1. In single-phase generation, how many electrical degrees are there in a single complete sinewave?   _C_
   a. 180°
   b. 90°
   c. 360°
   d. Depends on the RPM

2. When a conductor travels parallel to the magnetic lines of flux there is   _A_
   a. zero induced voltage.
   b. maximum induced voltage.
   c. 70.7 percent of maximum voltage.
   d. a small amount of residual voltage induced.

3. A revolving-field, stationary-armature generator is   _B_
   a. never used.
   b. used in large generation facilities.
   c. used only for demonstration models.
   d. not possible in generator construction.

4. A polyphase system is   _D_
   a. a three-phase system.
   b. a two-phase system.
   c. a six-phase system.
   d. two or more single-phase systems.

5. Three-phase alternator windings are displaced   _B_
   a. 90° apart.   c. 180° apart.
   b. 120° apart.   d. 360° apart.

6. The two-wattmeter power measurement method   _A_
   a. cannot be used in an unbalanced three-phase, four-wire system.
   b. is used on single phase only.
   c. can be used in an unbalanced three-phase, four-wire system.
   d. is used on single- and three-phase systems.

7. In a three-phase wye connection,   _B_
   a. line current = 1.73 × phase current.
   b. line current = phase current.
   c. line voltage = phase voltage.
   d. line voltage × 1.73 = phase voltage.

8. In a three-phase delta connection,   _D_
   a. line voltage = 1.73 × phase voltage.
   b. line current = phase current.
   c. line voltage × 1.73 = phase voltage.
   d. line voltage = phase voltage.

9. The wye connection is usually wired to a   _B_
   a. five-wire, three-phase line.   c. six-wire line.
   b. four-wire, three-phase line.   d. three-wire, single-phase line.

10. The wye connection is used in   _A_
    a. three-phase systems.   c. single-phase systems.
    b. two-phase systems.   d. special motor connections.

11. A three-phase, wye-connected generator winding has a phase   _A_
    current rating of 10 amperes. The line current rating is
    a. 10 amperes.   b. 30 amperes.   c. 17.3 amperes.   d. 15 amperes.

12. The voltage rating of an AC, wye-connected, three-phase   _C_
    generator is 208 volts. The voltage rating of each winding is
    a. 69.3 volts.   c. 120 volts.
    b. 208 volts.   d. 110 volts.

13. Six leads are brought out of a three-phase transformer. The leads   _A_
    are labeled A_1, A_2, B_1, B_2, C_1, and C_2. The wye connection
    can be made by connecting leads
    a. A_2, B_2, and C_2.
    b. A_2 to B_1, B_2 to C_1, and C_2 to A_1.
    c. A_1 to B_1 to C_1, and A_2 to B_2 to C_2.
    d. A_2 to B_1 and B_2 to C_1.

14. When the line voltage of a three-phase, four-wire system   _c_
    is 220 volts, the line to ground voltage will be

    a. 110 volts.   c. $\dfrac{220}{1.73}$ volts.

    b. 220 volts.   d. $\dfrac{220}{3}$ volts.

15. Six leads marked A_1, A_2, B_1, B_2, C_1, and C_2 are brought out of   _c_
    a three-phase transformer bank. If A_2 is connected to B_1, B_2 to
    C_1, and C_2 to A_1, the transformer windings are connected in
    a. wye.   c. delta.
    b. series.   d. parallel.

16. If the phase voltage of a delta-connected generator is 220   _c_
    volts, the line-rated voltage is
    a. 660 volts.   c. 220 volts.
    b. 330 volts.   d. 127 volts.

17. The delta connection is connected to a   _c_
    a. four-wire, three-phase line.
    b. five-wire, three-phase line.
    c. three-wire, three-phase line.
    d. six-wire, three-phase line.

18. An AC generator has delta connections. If each winding is   _c_
    rated at 20 amperes, the line rating is
    a. 60 amperes.   c. 34.6 amperes.
    b. 20 amperes.   d. 30 amperes.

19. A wye-connected AC generator is rated at 208 volts and 25   _B_
    amperes. The phase winding rating is
    a. 208 volts, 25 amperes.   c. 120 volts, 14.4 amperes.
    b. 120 volts, 25 amperes.   d. 208 volts, 14.4 amperes.

20. A delta-connected AC generator is rated at 220 volts and   _A_
    17.3 amperes. The phase winding rating is
    a. 220 volts.   c. 381 volts.
    b. 127 volts.   d. 660 volts.

21. A 220-V, 17.3-A, delta-connected AC generator is reconnected   _c_
    in wye. The new line voltage rating is
    a. 220 volts.   c. 381 volts.
    b. 127 volts.   d. 660 volts.

22. When three-phase windings are connected in delta, the
    coils are connected in

    _B_

    a. an open series circuit.   c. parallel.
    b. a closed series circuit.   d. series parallel.

23. In electrical terminology, the word "delta" means

    _C_

    a. a deposit at the mouth of a river.
    b. the Greek letter represented by a triangle.
    c. coils joined together at the ends.
    d. coils connected in an open series circuit.

11

PHYSICAL AND ELECTRICAL CHARACTERISTICS OF THREE-PHASE ALTERNATORS

OBJECTIVES

After studying this unit, the student should be able to

- describe the purpose of an alternator.
- describe the ways in which the field of an alternator is established and how the alternator operates.
- explain the operation of the field discharge circuit.
- state how the frequency of an alternator can be determined and give the formula for calculating the frequency.
- explain how voltage control for an alternator is accomplished.
- describe the structure and operation of a rotating-field alternator.
- diagram alternator connections.
- explain three-phase voltages.

An *alternator* is a machine designed to generate alternating current (AC). This machine is the major electrical unit in power plants.

The alternator converts the mechanical energy of a prime mover, such as a diesel engine, steam turbine, or water turbine, to electrical energy. Another prime mover that has become increasingly important in the generation of electricity is the power of wind (see Unit 13).

THREE-PHASE VOLTAGES

Three phase is the most common polyphase electrical system. *Poly* means more than one. It is, in this instance, a system having three distinct voltages that are out of step with one another. There are 120 electrical degrees between each voltage sine waveform. Figure 11-1 shows sinewave relationships taken on an electrical oscillograph instrument trace. This display shows the voltage relationships of the three-phases, 120 electrical degrees apart. This can be taken at any point in a three-phase system. The three phases are generated by placing each phase coil in the alternator 120° apart, mechanically. A rotating DC magnetic field will then cut each phase coil in succession, inducing a voltage in each armature coil. Armatures are the electrical components of the AC generator that have voltage induced into them. Armatures may be either the rotating piece of the alternator or the stationary component of the alternator. These armature coils may be connected internally or externally in a delta or a wye (star) connection. Rotating fields are more commonly used than stationary fields because generating large amounts of current would require larger sizes of conductors and iron to rotate. Therefore, it is more practical to make the armature stationary and rotate the magnetic field.

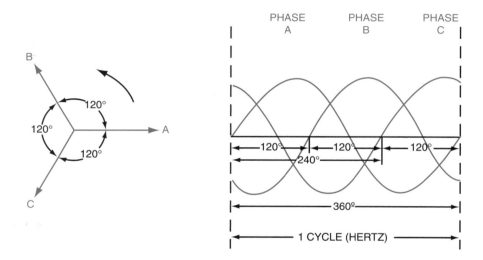

Figure 11-1 Electrical displacement and generation of a three-phase voltage.

Wye (star) and delta connections are shown in Figure 11-2. These connections are shown in more detail in Units 19 and 20.

ALTERNATOR TYPES

Two principal types of synchronous alternators are (1) the rotating-armature alternator and (2) the rotating-field alternator.

Figure 11-3 illustrates an alternator with a stationary field, a revolving armature, and the elementary wiring symbol for a three-phase alternator. The armature consists of the windings into which current is induced. The magnetic field for this type of alternator is established by a set of stationary field poles mounted on the periphery of the alternator frame. The field flux created by these poles is cut by conductors inserted in slots on the surface of the rotating armature. The armature conductors are arranged in a circuit that terminates in slip rings. Alternating current induced in the armature circuit is fed to the load circuit by carbon brushes that make contact with the slip rings.

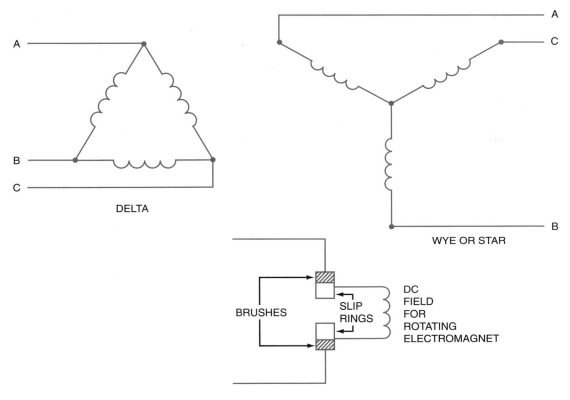

Figure 11-2 Three-phase internal generator connections and a stationary armature with a rotating DC field.

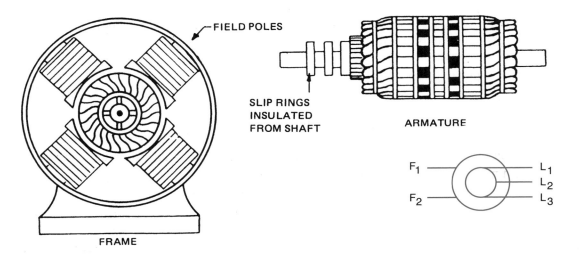

Figure 11-3 Parts of an alternator of the rotating-armature type.

The rotating-armature alternator generally is used for low-power installations. The fact that the load current must be conducted from the machine through a sliding contact at the slip rings poses many design problems at higher values of load current and voltage. One alternator design has semiconductor rectifier diodes installed on the exciter field, thus eliminating the brushes and slip rings for the revolving field alternator (see the section called "Brushless Exciters with Solid-State Voltage Control").

ROTATING-FIELD ALTERNATORS

Rotating-field alternators are used extensively because of the ease with which a high voltage or high current load can be connected to the alternator. The load connections do not need to draw power from the armature through brushes riding on slip rings. In this alternator, the output power is taken from the stationary armature through solid electrical connections. The armature is the stationary component of the generator, and the DC field is connected to the rotating field. The field windings in large generators are typically smaller than the armature windings. Therefore, the field has less mass and weight to rotate and can be supplied with a relatively small amount of DC to generate a relatively large amount of AC in the armature.

FIELD EXCITATION

DC must be used in the electromagnetic field circuit of an alternator. As a result, all types of alternators must be supplied with field current from a DC source, except for small permanent magnet fields. The DC source may be a DC generator operated on the same shaft as the alternator. In this case, the DC generator is called an *exciter,* shown on the self-excited synchronous alternator in Figure 11-4(A). The circuit diagram for this alternator is shown in Figure 11-4(B). In installations where a number of alternators require excitation power, this power is supplied by a DC generator driven by a separate prime mover. The output terminals of this generator connect to a *DC exciter bus* from which other alternators receive their excitation power by means of brushes and slip rings for the rotating-field type of alternator.

FIELD DISCHARGE CIRCUIT

A field discharge switch is used in the excitation circuit of an alternator. This switch eliminates the potential danger to personnel and equipment resulting from the high inductive voltage created when the DC field circuit is opened.

Figure 11-5 illustrates the connections for the field circuit of a separately excited alternator. With the discharge switch closed, the field circuit is energized and the field discharge switch functions as a normal double-pole, single-throw switch.

The discharge switch shown in Figure 11-6 has an auxiliary switch blade at A in addition to the normal blades at C and D (Figure 11-5).

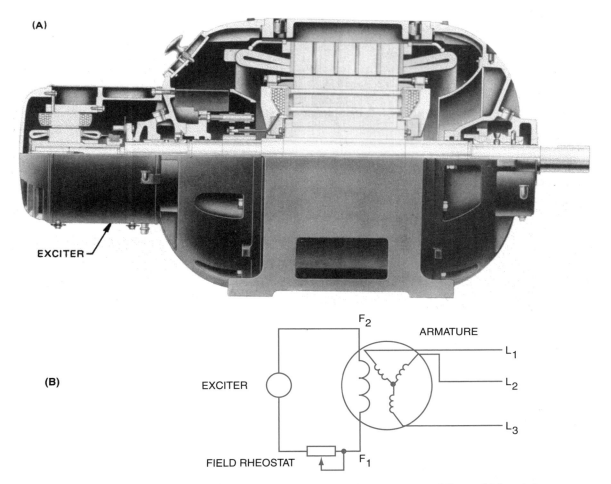

Figure 11-4 (A) Self-excited synchronous alternator. (*Photo courtesy of General Electric Company*) **(B) Circuit diagram.**

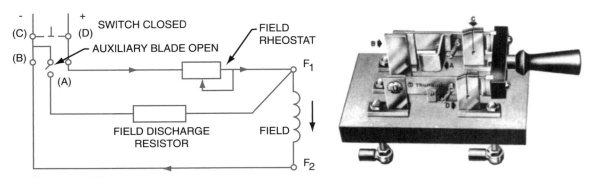

Figure 11-5 Field discharge circuit.

Figure 11-6 Field discharge switch.

When you want to open the field circuit, the following actions must take place.

- Before the main switch contacts open, switch blade A meets contact B and thus provides a second path for the current through the field discharge resistor.
- When the main switch contacts C-D open (shown in Figure 11-7), high inductive voltage is created in the field coils by the collapsing magnetic field.
- The high voltage is dissipated by sending a current through the field discharge resistor.

This procedure eliminates the possibility of damage to the insulation of the field windings as well as danger to anyone opening the circuit using a standard double-pole switch. A field discharge circuit is used with all types of separately excited alternators.

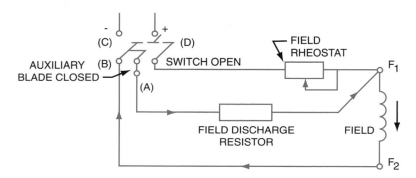

Figure 11-7 Field discharge circuit.

FREQUENCY

The frequency of an alternator is a direct function of (a) the speed of rotation of the armature or the field and (b) the number of poles in the field circuit. The frequency commonly used in the United States is 60 cycles per second or 60 Hz. Power companies are particularly concerned with maintaining a constant frequency for their energy output because many devices depend on a constant value of frequency. This constant value is achieved by sensitive control of the prime mover speed, driving the alternator.

If the number of field poles in a given alternator is known, then it is possible to determine the speed required to produce a desired frequency. One cycle of voltage is generated each time an armature conductor passes across two field poles of opposite magnetic polarity. The frequency in cycles per second or hertz is the number of pairs of poles passed by the conductor in a second. Because the speed of rotating machinery is given in revolutions per minute (r/min), the speed in revolutions per second is obtained by dividing the speed (r/min) by 60. In a two-pole alternator the frequency is

$$f = \frac{\text{number of poles}}{2} \times \frac{\text{rev/min}}{60}$$

$$600 = \frac{8}{2} \times \frac{RPM}{60} \qquad 600 = 4 \times \frac{RPM}{60} \qquad 100 = \frac{RPM}{60} \quad 4000\ RPM$$

or

$$f = \frac{poles \times RPM}{120}$$

Where f = frequency in hertz (formerly cycles per second)

p = number of poles

RPM = speed in revolutions per minute

120 = conversion factor

The formula for frequency can be rearranged so that the speed required to give a desired frequency can be obtained:

$$RPM = \frac{120 \times f}{P}$$

If a two-pole alternator is to be operated at a frequency of 60 Hz, the correct speed is obtained from the formula RPM = (120 × f)/p:

$$RPM = \frac{120 \times 60}{2} = 3,600 \text{ RPM}$$

For a four-pole alternator operated at a frequency of 60 Hz, the required speed is

$$S = \frac{120 \times 60}{4} = 1,800 \text{ RPM}$$

The two preceding examples illustrate the previous statement that the frequency of an alternator is a direct function of the speed of rotation and the number of poles in the alternator field circuit.

VOLTAGE CONTROL

The voltage output of an alternator increases as the speed of rotation accelerates, thus increasing the lines of force cut per second. If the field excitation increases, the magnetic fields are increased to the point of magnetic saturation of the field poles. Increasing the magnetic field strength increases output voltage.

For practical purposes, an alternator must be operated at a constant speed to maintain a fixed frequency. Thus, the only feasible method of controlling the voltage output is to vary the field excitation.

Field controls are used to vary the resistance of the total field circuit. This variation of resistance, in turn, changes the value of field current [Figure 11-4(B)].

- A low value of field current results in less flux and less induced voltage at a given speed.

- A high field current results in greater field flux and a higher induced voltage at a given speed.
- The value of flux at which the field poles saturate determines the maximum voltage obtainable at a fixed speed and frequency.

STATOR WINDINGS

Figure 11-8 illustrates the stator (stationary or nonmoving) windings of a rotating-field, three-phase alternator. The three-phase armature windings are embedded 120° from one another in the slots of a laminated steel core that is clamped securely to the alternator frame. Output leads from the stator emerge from the bottom of the stator and connect directly to the load circuit. Slip rings and brushes are not required in a stationary winding of this type. As a result, higher values of output voltage and current are possible. Standard values of voltage output for a rotating-field alternator are as high as 13,800 volts.

Figure 11-8 Stator winding of an alternator. (*Photo courtesy of General Electric Company*)

ROTATING FIELD

The rotating portion of a rotating-field alternator consists of field poles mounted on a shaft that is driven by the prime mover. The magnetic flux established by the rotating field poles cuts across the conductors of the stator winding to produce the induced output voltage of the stator.

The following comparison can be made between the rotating-armature alternator and the rotating-field alternator. In the rotating-armature alternator, the armature conductors cut the flux established by stationary field poles. For the rotating-field alternator, the motionless conductors of the stator winding are cut by the flux established by rotating field poles. In each case, an induced voltage is generated.

Figure 11-9 shows a salient field rotor for low-speed, three-phase alternators. For this type of rotor, the field poles protrude from the rotor support structure, which is of steel construction and commonly consists of a hub, spokes, and rim. This support

Figure 11-9 Alternator rotor, salient-field type. (*Photo courtesy of General Electric Company*)

structure is called a *spider.* Each of the field poles is bolted to the spider. The field poles may be dovetailed to the spider in some alternators to provide a better support for the poles against the effects of centrifugal force.

Figure 11-10 shows a nonsalient rotor. This type of rotor has a smooth cylindrical surface. The field poles (usually two or four) do not protrude above this smooth surface. Nonsalient rotors are used to decrease windage losses on high-speed alternators, improve balance, and reduce noise.

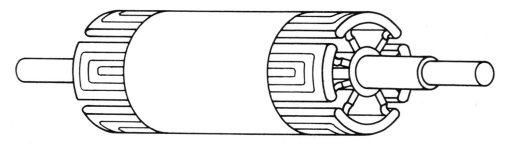

Figure 11-10 Alternator rotor, nonsalient-field type.

Power Supply for Rotor

The field windings of both salient and nonsalient rotors require DC power. Slip rings and brushes are used to feed the current to the windings at a potential of 100 to 250 volts DC. The brushes and rings are easily maintained because of the low values of field current encountered.

TERMINAL MARKINGS

A standard system of marking leads for field circuits has been established by ANSI (American National Standards Institute). The field leads for both alternators and generators are indicated by the markings F_1 and F_2. In addition, the F_1 lead always connects to the positive bus of the DC source. (See Figures 11-5 and 11-7.)

ALTERNATOR REGULATION

Regardless of the type of generator or alternator used in a system, the terminal output voltage of the machine varies with any change in the load current. The impedance of the windings and the power factor of the load circuit both influence the regulation of an alternator. An increase in load current in a pure resistive load circuit causes a decrease in output voltage. A voltage drop of approximately 10 percent is common when going from a condition of no load to full load in a typical alternator without voltage regulation controls.

For an inductive load, an increase in load current will cause a greater voltage drop than is obtained with a pure resistive load. A load with a low value of lagging power factor produces a large drop in output voltage.

A capacitive load circuit produces the opposite effect. In other words, the output voltage rises above the no-load value with an increase in load current and is high at a low value of leading power factor.

AUTOMATIC VOLTAGE CONTROL

Unlike DC generators, alternators cannot be compounded to alter the voltage-load characteristic. Moreover, output voltage variations are more likely to be severe because of changes in the load power factor. As a result, automatic voltage regulators generally are used with alternators.

Automatic voltage regulators change the alternator field current to compensate for any increase or decrease in the load voltage. A controller is used to increase or decrease the field current through the field circuit. As the AC line voltage falls, the field controller causes an increase in the flux and thus increases the induced voltage. An increase in the AC line voltage causes the controller to decrease the field current, flux, and induced voltage. Power companies stabilize line voltage by using a type of varying ratio transformer as a voltage regulator.

BRUSHLESS EXCITERS WITH SOLID-STATE VOLTAGE CONTROL

The permanent magnet generator (Figure 11-11) supplies high-frequency AC power input to the voltage regulator. Voltage and reactive current feedback information is provided to the regulator from potential and current transformers. Using these feedback signals and a reference point established by setting the voltage adjusting rheostat, the voltage regulator (which has a transfer switch allowing the operator to select automatic regulator control or manual control) provides a controlled DC output. The DC is fed to the field of the rotating exciter; the three-phase, high-frequency AC output is then rectified by a full-wave bridge. This rectified signal is applied to the main generator field. Fully rated, parallel, solid-state diodes with indicating fuses are provided to permit full-load generation with a diode (rectifier) out of service. The use of a stroboscope light permits the indicating fuses to be viewed during operation to determine if a diode has failed. Figure 11-12 shows a cutaway view of a brushless exciter. Figure 11-13 shows the rotating components of a brushless excitation system.

GENERATOR TROUBLESHOOTING AND MAINTENANCE

Test the insulation resistance of the generator at initial start-up and periodically track the insulation values.

Procedure for Testing Insulation Resistance

Disconnect all related components that have circuits connected to the generator windings. This includes the automatic voltage regulator, and any diodes on the rotor. If

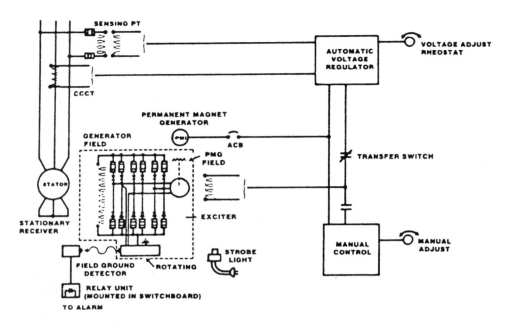

Figure 11-11 Diagram of an exciter with permanent magnet generator. (*Courtesy of Electric Machinery, Turbodyne Division, Dresser Industries, Inc.*)

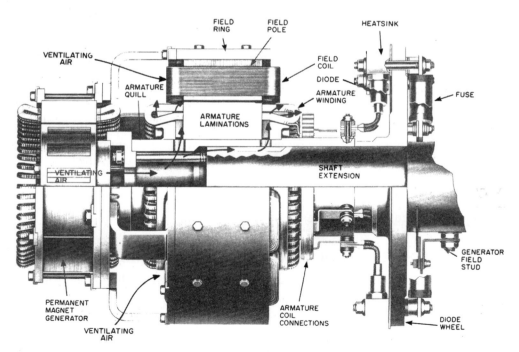

Figure 11-12 Cutaway view of brushless exciter showing components. (*Courtesy of Electric Machinery, Turbodyne Division, Dresser Industries, Inc.*)

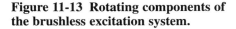

Figure 11-13 Rotating components of the brushless excitation system.

the generator has resistive temperature detectors (RTDs), disconnect or isolate them from the windings circuit. Use a megohmmeter to measure the resistance of the winding insulation. The measurements are made from winding to winding to make sure there is no electrical path between the coil windings. Also, the measurements are taken from the coils to the grounded frame to check for insulation breakdown. The resistance of the insulation will vary over time and may change if the windings are damp or in high humidity. However, the ratings should stay above 1.0 megohms to be considered sufficient insulation. An original reading of 100 megohms is typical when the machine is manufactured at the factory. If the generator arrives and the first readings are less than 1.0 megohm, the generator may have been damaged in shipment or the winding may be wet.

One method to dry the winding is to run the generator unloaded and with no field excitation. Run the generator for 30 minutes to allow the heat and ventilation of a running generator to dry the windings. Another method used to dry the windings and raise the insulation resistance (IR) is to open all venting covers and blow warm air over the damp windings. Be careful not to overheat the windings and actually burn the insulation. Continue checking the IR value as the windings dry out.

A third, more risky process is to actually have the generator generate a small voltage and current. This should be done only if other methods are not suitable. The process is to short the output terminals though a shorting conductor, large enough to hand the full-load current output of the generator. Disconnect the normal DC excitation leads and connect a separate DC source of adjustable low current at 24 V DC. Start the generator and then apply a very small excitation current. Very carefully increase the small excitation until the current being monitored at the output is at approximately 80 percent of the full-load current rating of the generator. A drying out curve as in Figure 11-14 should result if you take readings of the IR every 30 minutes.

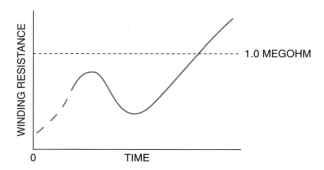

Figure 11-14 A typical curve that shows generator winding resistance as it dries out.

Other factors to check if the generator is running well mechanically but has problems with the power output are listed in the table.

No voltage at start-up	Check for residual voltage Check for proper speed Check for operation of automatic voltage regulator (AVR)
Unstable voltage when running	Check speed control Check AVR
Voltage higher than set	Speed too high Check AVR Check for capacitive load
Voltage at no-load lower than set	Check speed Check for residual voltage
Voltage with load lower than set	Check speed Check AVR

These are but a few of the common problems that can occur with electric generators. Each generator will have a troubleshooting manual with specific items to check if malfunction occurs.

SUMMARY

Three-phase alternators are similar to single-phase alternators in that they can generate electrical power through electromagnetic means. The rotating-field alternator is most common in large generating facilities. In electromechanical generation, a magnetic field is turned inside the housing that holds the circuit conductors. The speed of the rotating field is determined by the desired output frequency. The output voltage of a single generator will be controlled by the strength of the spinning electromagnet. By adjusting the amount of DC supplied to the spinning electromagnetic field, the level of AC output voltage can be controlled. Some generation systems do not connect DC power to the rotor using brushes and slip rings; instead, they use a system called a *brushless exciter* that supplies DC to the rotor through electromagnetic induction and rectifiers.

ACHIEVEMENT REVIEW

Select the correct answer for each of the following statements and place the corresponding letter in the space provided.

1. The armature of an alternator   _D_
   a. is the revolving member.
   b. is stationary.
   c. is the frame.
   d. consists of the windings into which the current is induced.

2. In alternators of the revolving-armature type,   _A_
   a. slip rings are required in the power output circuit.
   b. slip rings are required in the field circuit.
   c. slip rings are not required.
   d. one slip ring is required.

3. In a protective field discharge circuit, the auxiliary blade of   _C_
   the field switch inserts the discharge resistor
   a. at the instant the field circuit opens.
   b. immediately after the main blade loses contact.
   c. immediately before the main blade loses contact.
   d. immediately after the main blades make contact.

4. A field discharge circuit resistor   _D_
   a. is installed to stabilize line voltage.
   b. is installed to stabilize line current.
   c. improves regulation.
   d. eliminates danger to people and equipment.

5. The frequency of the alternator output   *A*
   a. is directly proportional to its speed.
   b. is inversely proportional to its speed.
   c. depends upon its field strength.
   d. is inversely proportional to the number of poles.

6. The speed of a six-pole, 60-Hz alternator is   *B*
   a. 600 r/min.   c. 1,800 r/min.
   b. 1,200 r/min.   d. 3,600 r/min.

7. To deliver power at a frequency of 400 Hz, an eight-pole   *C*
   alternator must be driven at what speed?
   a. 600 r/min.   c. 6,000 r/min.
   b. 3,600 r/min.   d. 8,000 r/min.

8. High-speed alternators are designed with   *D*
   a. a revolving armature and a nonsalient rotor.
   b. a revolving armature and a salient rotor.
   c. revolving fields and a salient rotor.
   d. revolving fields and a nonsalient rotor.

9. Changing the driven speed of an alternator   *D*
   a. changes the voltage magnitude to field saturation.
   b. changes the frequency output.
   c. does not affect voltage or frequency.
   d. both a and b are correct

10. The magnitude of the voltage output of an alternator is   *C*
    generally controlled by
    a. the speed of the prime mover.
    b. a field rheostat.
    c. variable resistance in the output lines.
    d. changing the power factor of the load.

11. Alternators use all but which one of the following systems to *DB*
    obtain field excitation?
    a. a separate DC power supply
    b. a self-excited AC field circuit
    c. a DC exciter on the same shaft as the alternator
    d. a rectifier to convert the output voltage for use in the field circuit

12. The greatest drop in output voltage results from taking full-load   _____*C*_____
    power from an alternator at a
    a. unity-power factor load.
    b. high-power factor capacitive load.
    c. low-power factor inductive load.
    d. medium-power factor capacitive load.

13. Three-phase voltage is   _____*B*_____
    a. three polyphase circuits.   c. three delta connections.
    b. three distinct voltages.   d. three wye connections.

14. The schematic diagram for a three-phase alternator is A, B, C, or D as shown in
    Figure 11-15. _____*A*_____

    _____

    _____

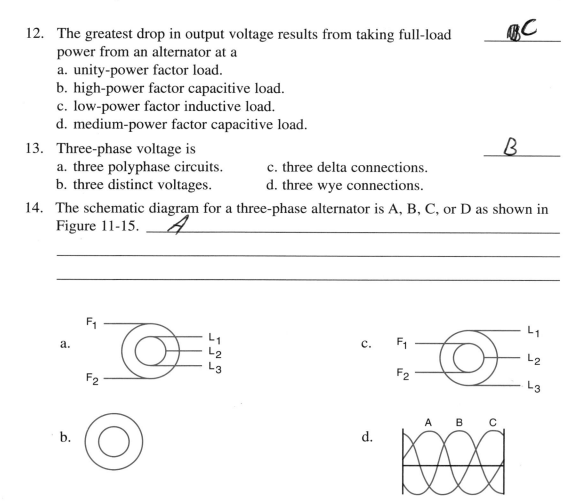

Figure 11-15 Schematic symbol for a three-phase alternator.

12

ENGINE-DRIVEN
GENERATING SETS

OBJECTIVES

After studying this unit, the student should be able to

- describe the purposes of engine-driven generating sets.
- list the advantages of using cogenerating sets.
- describe the operation of an automatic transfer switch.
- connect an automatic transfer switch.
- state *National Electrical Code*® requirements.

ENGINE-DRIVEN GENERATING SETS

Diesel, gasoline, or natural gas engine-driven generators are most commonly used to provide another source of emergency or standby power when normal utility power fails. Turbine power generator sets are also used in this application.

Sturdy, diesel-engine–powered generators are still popular in remote area power systems. The use of hybrid systems using natural energy, such as the wind and the sun, is growing dramatically. Although diesel generators will not become obsolete for remote site electrical power, a need will exist for alternate sources, as the role of the diesel-engine generator changes from a primary energy source to part of a combined source.

Most engine-driven generator sets are rated from a few hundred watts to several hundred kilowatts, although units rated as high as 3,000 kW have been successfully applied. Multiple units, with some working in parallel, are becoming more commonly used to increase generating capacity. Controls may be manual, remote, or automatic, depending on their application.

Transfer Switches

Elementary transfer switches are typically classified as "break before make" switching actions. The connection to the normal source of supply is broken before the connection is made to the emergency source. This action is intended to prevent the emergency generator from supplying power to the normal utility source. Likewise, as the switch transfers back to normal, the emergency source connection is broken before the main power is reconnected. This process is referred to as open transition. This open transition is all right for many loads as the automatic transfer takes less than 10 cycles or one-sixth of a second.

Closed transition is needed with some loads. In this situation, when the load is to be transferred back to the normal source, a synchronizing action must occur to synchronize the two sources so they can both be connected at the same time. The time that both sources are connected is very short, as short as 100 milliseconds, but the transfer is made without interrupting the power to the load as it transfers back to normal. The first transfer, from normal to emergency, is still open transition.

Another variation on the transfer of load is referred to as "soft loading," meaning the load is added gradually, or softly, to the auxiliary generator to enable loading of the on-site generation to meet the needs. Different modes of operation allow the soft loading to gradually transfer the load to the on-site generator to help reduce demand from the utility. The on-site generator can be used for cogeneration to sell power back to the utility and the load on the generator can be adjusted or programmed to stay constant to supply on-site demand as well as sell surplus power to the utility. This type of transfer is designed to have two sources connected simultaneously and the electrical load shifts from one to the other in a smooth, predictable fashion.

Depending on where the on-site generator is located, inside the building or separate from the building, the transfer switch can be located to efficiently serve the main normal power and the generator power feeds. This means that the switch may be located outdoors in an enclosure suited for the need or it may be installed indoors in a suitable enclosure. Figure 12-1 shows an automatic transfer switch.

An elementary diagram of a typical automatic transfer switch is shown in Figure 12-2. (The figure does not include engine starting and other controls.)

When the normal supply on the left side is energized, current flows from L_1 through the time-delay coil (TD) and back to L_2. After a predetermined setting of time delay in closing contact, the relay R coil becomes energized. Contact R then closes, and energizes the N coil. Power contacts N then close, supplying the load from the normal or preferred source. When the R coil is energized, it also opens the normally closed R contact interlock in the E coil emergency circuit. This safe action ensures that each power supply operates independently of the other.

When the normal power fails, all coils on the left, or normal supply side, become deenergized. Relay contact R drops to its normally closed position in the E (emergency coil) circuit. Coil E is then energized, which closes the E power contacts feeding the load from an emergency electrical supply.

The time delay action helps to ensure that the normal service does not supply the load intermittently with the emergency supply. In other words, the switch will wait a preset time until the normal supply is firmly established, before the load is reconnected to normal.

Figure 12-1 An automatic transfer switch transfers the power source for the electrical load from a normal supply to an alternate supply.

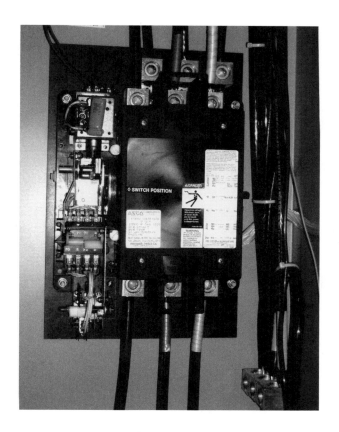

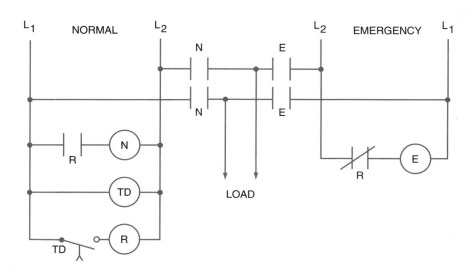

Figure 12-2 Elementary diagram of an automatic transfer switch.

EMERGENCY SYSTEMS

Applicable *National Electrical Code*® (*NEC*®) and local code rules are considered when an on-site generator is selected. These rules differ, depending on whether the generating set is to function as a power source in a health care facility, such as a hospital, a standby power system, or as an emergency system.

On-site generator systems generally are installed wherever great numbers of people gather, and where artificial lighting is required, such as in hotels, theaters, sports arenas, hospitals, and similar institutions. In addition to lighting, emergency systems supply loads that are essential to life and safety. Such installations include fire pumps, ventilation, refrigeration, and signaling systems. (Refer to *Article 700* of the *National Electrical Code*®.)

Optional Standby Power Generation Systems

Optional standby power generation systems include alternate power systems for applications such as heating, refrigeration, data processing, or communication systems where interruption of normal power would cause human discomfort or damage to a product in manufacturing, but where life and safety do not depend on the system. (Refer to *Article 702* of the *NEC*®.)

UNINTERRUPTIBLE POWER SUPPLIES

Uninterruptible power supplies (UPSs) are used as power systems to supply critical electronic equipment, including electronic computer and data processing equipment as described in *Article 645—Information Technology Equipment* of the *National Electrical Code*®.

The basic system includes an electronic section that takes the normal line power and converts it to DC power to act as a battery charger. The system has a set of batteries that are constantly being charged. These batteries are also connected to an electronic inverter system to change the DC battery power back to normal AC power to supply an AC load. The concept is to constantly have battery power connected and available to provide a standard level of power to the electronic equipment at all times. The battery system is sized to provide the needed volt-amp capacity to the system for a specified period of time. The UPS system also provides protection from low-voltage conditions (brown-outs) and momentary surges or power delivery failures. These systems can provide continuous, filtered, and regulated power to sensitive equipment. The systems themselves can be sized to provide for a few computers or to supply hundreds of computers or other essential equipment. Figure 12-3 shows UPS control panels. If the UPS system is large and requires a bank of batteries to supply the uninterruptible power, the batteries must be installed according to the *National Electrical Code*® *Article 480*.

Figure 12-3 The readout monitor panel for a UPS system indicates current status.

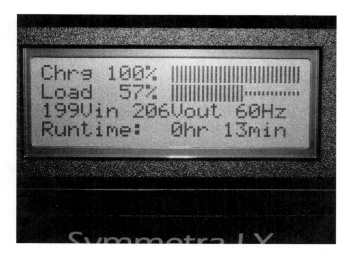

Health Care Facilities

Health care facilities are governed by several *National Electrical Code*® rules concerning power sources, emergency systems, and essential electrical systems. In particular, refer to *NEC*®*, Article 517*. *Article 700* refers to emergency systems, and *Article 701* refers to legally required standby systems.

Figure 12-4 shows a diesel-driven emergency power system consisting of four 450-kW electric generating sets. The system is electronically synchronized to deliver 1.8 million watts of emergency power for a hospital. Each unit can also be operated independently of the other units. Figure 12-5 shows the digital control panel for an alternator.

LEGALLY REQUIRED STANDBY SYSTEMS

NEC® *Article 701* states that legally required standby power systems are those systems required by municipal, state, federal, or other codes or a government agency having jurisdiction. In the event of failure of the normal power source, these systems are intended to take over automatically. Legally required standby power systems are installed to serve such loads as communication systems, ventilation and smoke removal systems, sewage disposal, and rescue and fire fighting equipment, among others. These systems *must* be installed within the guidelines of the authority having jurisdiction.

COGENERATING PLANTS

Cogeneration is being used to help reduce the cost of purchasing power from a local utility. Many forms of cogeneration are available. Some use the concept of recovering the energy from some manufacturing process to drive electrical generators on site.

Figure 12-4 1.8 million-watt, diesel-driven emergency power generating system.

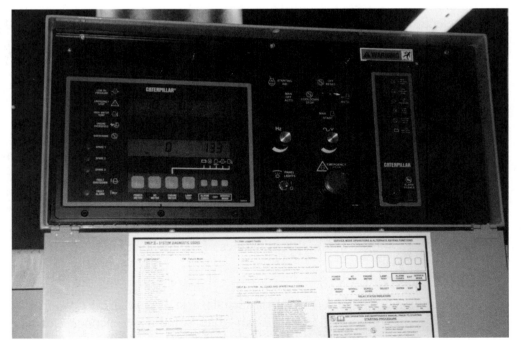

Figure 12-5 Digital control panel for an alternator.

Some cogenerating plants are diesel-powered electric generators that are designed to recapture and use the waste heat from both their exhaust and cooling systems. Figure 12-6 shows a standby generator with transfer switch and controller cabinet.

Although cogenerating plants are not a new concept, they are now being used to combat rising prices charged by public utility companies for power generation. The nation's largest manufacturers of diesel engines provide competition for the electric utility companies in the United States. As a result, these manufacturers have been concentrating on selling cogenerating plants.

Alongside cogeneration of electrical power is the deregulation of the electrical utility monopoly. This deregulation of the power generation and delivery system allows many power producers to supply power to the national power grid. In turn, customers can select the power producer that can supply the cheapest and most reliable power. In some cases, this means that power producers are many states distant and the local utility system provides the conductors to deliver the power. Many customers are installing their own on-site power generation to ensure that their own power supply is not interrupted because of system overloads or other problems.

Equipped with cogenerating plants, energy users need no longer rely on public utilities because users can make their own electricity at a lower cost and provide heating and cooling for their buildings.

Figure 12-6 A diesel-powered standby/peak-shaving plant.

Various technical methods have been devised for using cogenerating plants. However, they all capitalize on the fact that the generation of electricity wastes about twice as much energy in the form of heat as the amount of energy that can be generated as electricity. Steam heat, as a waste byproduct of manufacturing processes, is now harnessed and used to turn steam turbine electric alternators. This electricity, when not needed, may be sold to the public utility that services the plant.

As the energy-saving application of cogeneration results in greater demand for electrical work, there are more jobs for electricians. It also should create a particular need for power generator operators with the skills to install, operate, and maintain cogenerating equipment.

SUMMARY

When the normal source of electrical power is interrupted, businesses and other organizations may require immediate restoration of power to continue critical operations or to sustain life and safe conditions. Several methods can be used to provide power, and different criteria determine which system is required or the best to use. Use the *National Electrical Code*® to determine which system is appropriate to install. In addition to required systems, large power consumers may generate their own power on site as part of money-saving activities. This cogeneration is often used to reduce the amount of energy purchased from the utility or to supply the high energy peaks in a facility to reduce demand charges by the utility.

ACHIEVEMENT REVIEW

Select the correct answer for each of the following statements and place the corresponding letter in the space provided.

1. Engine-driven generating sets are used for   _____
    a. emergency systems.   c. cogenerating plants.
    b. standby power.   d. all of the above

2. With an automatic transfer switch, as shown in Figure 12-1,   _____
    how does the emergency supply feed the load when power fails?
    a. TD energizes R.
    b. Normally open contact R opens.
    c. Normally closed contact R closes.
    d. Power contacts N close.

3. Generating capacities may be increased by using   _____
    a. parallel multiple units.   c. turbines.
    b. series multiple units.   d. diesels.

4. Cogenerating sets are used
   a. to supply emergency power.   c. to conserve energy.
   b. to supply standby power.   d. in health care facilities.
   _____

5. Electrical capacity is gained with several small generating sets by
   a. paralleling machines on the line.
   b. reducing the load.
   c. placing machines on the line in series.
   d. none of the above
   _____

U • N • I • T
13

ALTERNATIVE POWER SOURCES

OBJECTIVES

After studying this unit, the student should be able to

- describe various types of alternative local power sources.
- determine the type of power generation to best fit a particular need.
- list the advantages and disadvantages of different systems.
- describe the connection methods needed to connect power systems to a load.
- list the *National Electrical Code*® articles that pertain to special equipment.

POWER SOURCES

Wind, fuel cells, microturbine, solar photovoltaic power, and flywheel reserve power are all alternate sources of electric power. These systems are relatively small sources compared to coal or nuclear plants. They produce much less power than a hydro-electric generating station and often less than oil-fired steam generation or diesel peaking plants. These small generation methods are often used to reduce the demand on the power grid that is normally supplied from utility power stations. Sometimes these sources are used in remote locations where the power grid is weak or nonexistent. Very often these sources are used to power specific applications, reduce the power needs from the utility, or supplement the utility supply capacity.

Wind Power

As more energy sources are sought, the trend is to convert other sources of energy (such as wind) to electric power and then distribute the electricity to where it is needed.

Wind-powered generators are becoming increasing popular for sources of power with no direct charge for the source of energy. Unlike the workhorses of commercial generation, such as coal and nuclear power, where the procurement, transportation, storage, and waste are problems that must be dealt with, wind is free of these drawbacks. The third major source of power (after coal and nuclear power) is water or hydroelectric generation, in which the resource is free and available as long as it rains. Hydroelectric power generation requires a source of water supply, a topography with a large vertical drop for falling water, and storage of the potential energy behind hydroelectric dams, which creates its own sets of problems.

Generating Power from Wind

The commercial generation of power derived from the wind is a simple concept, but the technology has improved greatly in the recent past making it more feasible for economical production. Wind-generation "farms" are built after careful studies indicate that enough consistent wind is available to make an investment in the towers, turbines, generators, and transformers needed to provide power for the power grid. These "farms" have many towers lined up like "crops" to harvest the wind.

Wind power has the benefits of cheap, renewable energy, and no waste or storage problems. The drawback is, of course, that the wind must blow to produce electricity. In many parts of the world and in the United States, some regions have significant wind power available on a consistent basis, which makes them good candidates for wind-powered generation. For the most part, these systems supplement the base load generators providing energy locally or to the power grid as "extra" electric power so the base load can reduce its power output. The current technology is to drive induction generators at synchronous speed, usually 3,600 RPM, to produce 60-Hz output power for the United States. To do this, the variable pitch wind turbine blades are computer controlled

Figure 13-1 Wind-generation farms converting wind energy to electrical energy.

to face into the wind and spin at a precise speed. The 70–100-ft long blades collect a great amount of energy from the wind. See Figure 13-1 for a 100-ft tower and blades that create a 140-ft diameter circle. Offshore wind generators are now using rotors that are 110 meters (approx 360 ft) in diameter. The wind turbine will spin at the same speed but with more or less torque, depending on the force of the wind. The mechanical output from the spinning turbine is fed through a transmission box to increase the speed of the drive shaft to the desired synchronous speed. The generator is mounted to the transmission output and spins a magnetic field inside the fixed armature to generate power, as a typical generator would. See Figure 13-2 for a simplified view. Commercial generators range in size from 75 kW to 1.8 mW. The output power is matched to the power transmission system through automatic synchronization and paralleling switchgear, much the same as diesel cogeneration or standby systems. The wind-powered generator will

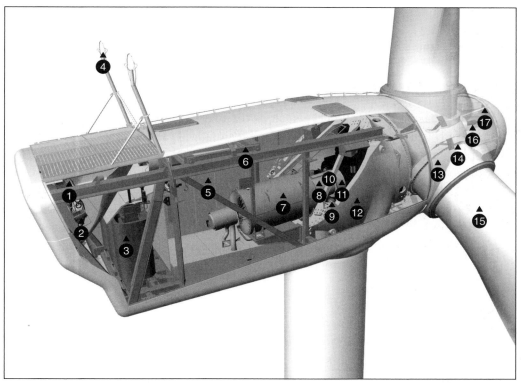

1 Oil Cooler	7 Variable Speed Generator	13 Blade Bearing
2 Generator Cooler	8 Composite Disc Coupling	14 Blade Hub
3 Transformer	9 Yaw Gears	15 Blade
4 Ultrasonic Wind Sensors	10 Gearbox	16 Pitch Cylinder
5 VMP-Top Controller with Converter	11 Parking Brake	17 Hub Controller
6 Service Crane	12 Machine Foundation	

Figure 13-2 Cutaway view of a modern wind-powered generator with a transformer in the nacelle (enclosure).

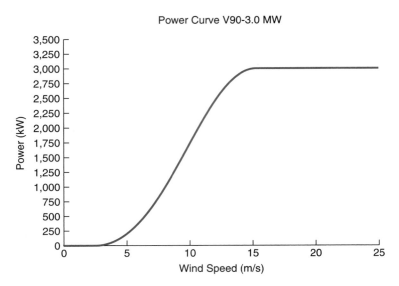

Figure 13-3 A power curve for a 3.0-mW generator. Wind speed in m/s, output in kW.

provide power based on a load curve such as the one in Figure 13-3. The turbines have computer controls to take advantage of the wind situations, and a minimum wind threshold will produce output power. Wind power below that threshold does not provide enough power for the generators and is taken off-line. As the wind becomes more powerful, there is a maximum wind speed that can produce power without damage to the turbine. Above that speed, the turbine will shut down and lock the blades to prevent damage to the blades or the tower. As each generator in the wind farm produces power, it allows a base load plant to reduce its power output and save money on the resources consumed.

Fuel Cells

Fuel cells generate electrical power through a chemical reaction. Fuel cells are becoming more abundant and more economical as an alternative energy solution. Fuel cells have been used for years for small-scale electrical power needs and have been used in vehicles ranging from the space shuttle to city buses, trucks, and passenger cars. Different types of fuel cells exist that produce DC electric power by using different types of electrolytes and different compositions of cathode and anode.

The basic concept of all fuel cells is the same in that they use hydrogen as a fuel. Hydrogen is plentiful but needs to be processed for fuel cell use. As hydrogen enters the fuel cell at the anode, a chemical reaction occurs that pulls the electron away from the

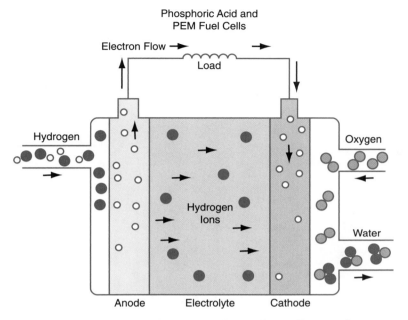

Figure 13-4 A proton exchange membrane fuel cell operation.

hydrogen atom. These stripped electrons are the electron current source for the external electrical circuit. The positive hydrogen ions are now in the electrolyte (see Figure 13-4). Another resource that is needed to complete the chemical reaction is oxygen, which enters the cathode. The oxygen is taken from the surrounding air. As the electrons return from the electric circuit, they recombine with the hydrogen ions and the oxygen atoms to create H_2O, or water. The waste from this reaction is water, which is sent down a drain. The other byproduct of the reaction is heat. This heat is often captured and used for other needs such as water heating or space heating, or even cooling using absorption-type chillers, making the fuel cell very efficient. The process of using the electrical DC output for customer use usually requires that the power be inverted through an electronic system much like the frequency-producing components of a variable frequency drive. See Figure 13-5 for an operating fuel cell producing 5 kW of power for a customer site.

Types of Fuel Cells

Different types of fuel cells have specific advantages and disadvantages. *Alkaline fuel cells* use potassium hydroxide in water as the electrolyte. They are about 70 percent efficient and operate at about 300 to 400° Fahrenheit. These cells have small electrical outputs, from 300 W to 1000 kW; however, they need pure hydrogen fuel and pure oxygen, and have expensive platinum electrodes. *Molten carbonate* fuel cells use

Figure 13-5 A fuel cell installation powering a library building.

high-temperature salts for an electrolyte and can leak. These fuel cells are 60–80 percent efficient and can produce up to 2 mW. They use an inexpensive nickel electrode. The major drawback is the high heat of approximately 1200°F. Because it takes time to reach operating temperature, the cell is best used for continuous load under supervision because of the high heat. *Phosphoric acid* cells have a lower operating temperature—300–400°F—and have an efficiency of about 60 percent. The electrolyte is phosphoric acid, which allows the cell to use less concentrated fuels and come in sizes from 200 kW to 11 mW. *Proton exchange membrane (PEM)* fuel cells are about 45 percent efficient and operate at about 174°F. The electrolyte is a solid membrane that does not leak, which makes PEMs suitable for mobile units between 50 W and 200 kW. The hydrogen fuel must be purified and the electrodes are expensive platinum.

NEC® Article 692 Fuel Cell Systems

Article 692.2 provides the definitions needed to determine the *NEC®* requirements for fuel cell installations. Because fuel cells are often supplementary power sources, the *Code*® requires that a permanent plaque or directory that lists all the power sources be established at each service location. The fuel cell must be evaluated and listed for the intended purpose, and the listed current on the nameplate shall be the circuit current rating. The feeder conductors that connect the fuel cell to the premises wiring must be sized on the nameplate or the rating of the integral overcurrent protection on the fuel cell.

Article 692.10 (C) states that a fuel cell that only supplies a 120-volt single circuit must be posted with a sign that warns that no multiwire branch circuits can be attached. *NEC*® Chapter 3—*Wiring Methods* are permitted to be used with the fuel cell installations as well as any specific equipment listed for fuel cells. *Article 692.41 (B)* dictates how to ground systems that are not "stand-alone" systems.

When the fuel cell is used to feed power back to a supply grid, the fuel cell must be able to detect when the utility power is lost. If the supply is lost, the fuel cell must not feed power back to the supply grid. This is a safety feature for utility line workers who would assume the power is off and do not want power to backfeed on to their lines.

Microturbine

Microturbines are another source of local power that can be used for stand-alone power at a site or as cogeneration that is supplied back to a power grid. As the name implies, microturbines are generators driven by mechanical turbines that are small scale compared to the large commercial generators. The microturbine consists of a system with a compressor, a combustor, a turbine, and a permanent magnet generator. The turbine is designed to spin at a very high speed: 96,000 RPM. This turns the rotor with a mounted two-pole permanent magnet inside the fixed armature. The generator actually produces a variable frequency, up to 1,600-Hz AC output voltage, which is rectified and fed to a DC bus section of the output. The power is then inverted through an IGBET inverter to produce a typical output of three-phase, 480-volt, 60-Hz output. Microturbines are typically 30 kW–60 kW each, so larger loads would need two or more systems connected via paralleling gear.

Microturbines are meant to run continuously for up to a year, then are removed from service for minor maintenance, and then restarted. These systems are designed to run on "air bearings," so there is very little friction and no need for bearing replacement or lubrication, or coolants. They can be fueled by natural gas, propane, diesel, or kerosene. They can even use "waste gas" as a fuel. Waste gas may be methane from sewage treatment, landfill venting, agricultural digesters, etc.

Solar Photovoltaic

Photovoltaic (PV) systems are made in two different ways. They can be produced in discrete cell technology or in integrated thin film technology. Discrete cell technology uses single crystal silicon wafers. Commercial single wafers are approximately

15 percent efficient, meaning that 15 percent of the amount of solar energy is converted to equivalent electrical energy. Multicrystal silicon is another form of discrete cell technology. These wafers come from solid blocks of cast silicon. They are less expensive to manufacture, but also slightly less efficient. Dedritic web is the most efficient form of the discrete cell technology. These are the most expensive and are used in high reliability and high efficiency (up to 30 percent) areas such as power for space systems.

Integrated thin film technology uses copper indium diselenide (CIS). A version of this thin film called amorphous silicon (a-Si) is currently used in calculators and watches. This is relatively inefficient and is not used much for large power needs.

Power systems can be used on a distributed power basis, meaning they can provide electrical power where needed, not only within wiring distance of a power grid. They do not have moving parts, so they require very little maintenance. The cells are not consumed, so they will last for decades. Unfortunately, they can only provide electrical power when the sun is shining, so the systems most often require some form of electrical storage system such as batteries. Because batteries store DC, the system usually has an inverter to convert the DC to useable AC.

NEC® Article 690

Article 690.1 refers to the scope of the article, the producing of electricity through direct photovoltaic means. This is rather than by heating water from solar energy, and then turning a turbine to produce electricity. These systems may be stand-alone or interconnected to other electrical sources. They may produce DC or AC and either have battery storage or not. In the 2008 *NEC®*, there are several diagrammatic examples of each system. *Article 690.2* has extensive definitions that pertain to the photovoltaic system.

General installation requirements are contained in *Article 690.4*. Generally the output from the photocells cannot be on conductors mixed with the branch circuit conductors or feeders from another system. When installing a system, be careful to connect the grounded conductors so that the continuity cannot be disconnected by the removal of a system component. If solar panels are roof mounted on dwellings, they must be protected by DC ground fault protection, which interrupts the fault current and provides an indication of the fault. If the grounded conductor is interrupted to clear the fault, the ungrounded conductor must also be opened.

Micro-Hydro Generators

Another alternate source of electrical energy is the use of small electrical generators that use the force of flowing water to provide the mechanical energy needed to spin the rotor. The concept is to generate small amounts of power by directing water to these generators from moving streams, or even from moving water such as tidal pools, or from moving boats. These small generators are used to provide supplementary power to remote locations or additional power without the need for fuels. There are three basic types of turbines currently designed for different applications.

The impulse style uses a "spinner"-type propeller where the water is channeled to the propeller and strikes it to cause the mechanical force. This system works well if there is vertical drop or "head" of 20 feet or more but requires little water volume. Another style of microturbines is a reaction-style turbine. With this style there must be a large flow of water directed to the impellor, but it does not need much head. It can be installed in a moving stream. The third style is a submersible impeller that takes advantage of a fast-moving stream or is used over the side of a moving boat. All these systems are designed to charge batteries at typically 12 to 24 V DC. The batteries are part of a storage system that can receive power from multiple sources. The output is then connected to inverters to provide the system voltage desired.

UPS Using Flywheel Technology

Uninterruptible power supply (UPS) is used where there cannot be an interruption to the power supply. (See UPS battery systems in Unit 12.) At least one manufacturer has incorporated the physics of momentum into a power generation system. This system is called the battery-free uninterruptible power system (BF-UPS).

This alternative energy storage system uses the concept of stored energy in the inertia of a spinning flywheel. During normal power supply to a facility, the BF-UPS allows a large flywheel to spin at a constant speed. The flywheel is constantly spinning on high quality bearing, so the energy needed to maintain momentum in the flywheel is minimal. In the event of a line disturbance or a power failure, the spinning flywheel gives up its energy to keep an integral generator running with no interruption of power. The flywheel is sized to provide up to 2 minutes of full power before the flywheel begins to slow down. This is usually enough time to start a standby generator and bring it up to synchronization speed to pick up the load. A side benefit of the flywheel system is that it can actually supply the DC starting power the generator needs, so there are no failed batteries for the load power or the generator starting system. There is no battery charging or battery maintenance, so the environmental considerations are eliminated. Because no batteries are required, much less floor space is dedicated to this type of UPS. The drawback of the system is that it is usually sized to provide load power for a short duration, just enough to get a traditional generator operating.

The installation of this system has the same parameters as a normal standby/UPS system but without the need for batteries.

SUMMARY

Besides the usual sources of electrical power from the utility power grid, there are alternate sources for spot supply or remote supply of electricity. UPS battery systems and engine-driven generating sets were discussed in Unit 12, but other unique systems are now becoming more popular and are more cost effective than before. Each system has its own benefits and drawbacks, but each of these alternative systems other than micro-turbines, does not burn fossil fuels or create the emission problems of other fuels. Nuclear fuel does not create the air pollution problems, but has a lot of other inherent problems. Hydroelectric generation has a different set of problems, such as it can only be produced with the correct topographical conditions. Because these alternative sources are small-scale production, they are typically for local use. However, wind power is now tied to the power grid to help meet the need for power production on a regional basis.

ACHIEVEMENT REVIEW

Select the correct answer for each of the following statements and place the corresponding letter in the space provided.

1. Compared to a hydroelectric generating facility, wind power has a _____
   free source for fuel, but does not have the drawback of
   a. storage of potential energy.
   b. electromechanical generation from a generator.
   c. a need for specific location to access a renewable resource.
   d. a very large investment in facilities.

2. Wind-powered generators are designed   _____
   a. to have variable speed generators.
   b. to produce DC as an output for battery storage.
   c. to operate either at full capacity or zero output.
   d. to run at a synchronous speed at varying output power.

3. Fuel cells actually produce power through   _____
   a. electromagnetic processes.
   b. chemical reactions with hydrogen.
   c. electrolyses using nitrogen from the air.
   d. the burning of waste fuels.

4. Besides output power and water, the other byproduct   _____
   for a fuel cell is
   a. heat.
   b. waste hydrogen fuel rods.
   c. carbon dioxide.
   d. ionized air.

5. The code article for fuel cells states that   _____
   a. fuel cells cannot be connected to the power grid.
   b. fuel cells must not feed power to a dead power grid.
   c. fuel cells are for single phase power only.
   d. fuel cells are for DC power only.

6. Microturbines are meant for   _____
    a. intermittent operation.
    b. continuous operation.
    c. standby power only.
    d. low-speed, low-frequency operation.

7. PV systems are designed to   _____
    a. convert sunlight directly to DC voltage and current.
    b. create steam that turns a turbine.
    c. convert solar power directly to AC voltage and current.
    d. be "stand-alone" systems only.

8. A UPS system without batteries uses _____ to store energy.   _____
    a. capacitors.
    b. lead acid cells.
    c. a flywheel diode.
    d. a mechanical flywheel.

14

PARALLEL OPERATION OF THREE-PHASE ALTERNATORS

OBJECTIVES

After studying this unit, the student should be able to

- state the conditions that require two alternators to be paralleled.
- describe the use of synchronizing lamps in the three dark method and the two bright, one dark method of synchronizing alternators.
- demonstrate the procedure for paralleling two three-phase alternators.
- state the effect of changes in field excitation and speed on the division of load between paralleled alternators.
- describe "reverse power."

WHEN TO PARALLEL ALTERNATORS

Alternators are paralleled for the same reasons that make it necessary to parallel DC generators. Two alternators are paralleled whenever the power demand of the load circuit is greater than the power output of a single alternator.

When DC generators are paralleled, the output voltage and electrical polarity of the machines must be matched with the voltage and polarity of the line. The same matching is required when alternators are paralleled. However, the matching of alternator polarity to that of the line presents problems not encountered when matching DC generator and line polarities. The output voltage of an alternator is continuously changing in both magnitude and polarity at a definite frequency. Thus, when two alternators are paralleled, not only must the rate of the rise and fall of voltage in both alternators be equal, but the rise and fall of voltage in one machine must be exactly in step with the rise and fall of voltage in the other machine. When two alternators are in step, they are said to be in *synchronism*. Alternators cannot be paralleled until their voltages, frequencies, and instantaneous polarities are nearly equal.

Figure 14-1 shows a comparison of the voltage curves of one of the phases of two three-phase generators operating independently but at different frequencies. The voltage curves must be in synchronism before paralleling machines.

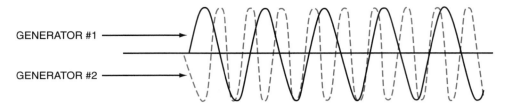

GENERATOR #1

GENERATOR #2

Figure 14-1 Periodic time relationship of the out-of-phase voltages of two generators running at different frequencies.

The output voltage of an alternator can be controlled by varying the strength of the direct current in the field circuit of the alternator. A field current control can be used to vary the DC current. Because the frequency of an alternator varies directly with speed changes, you must be able to control the speed of at least one alternator in an installation containing two machines.

ACHIEVING SYNCHRONIZATION

To synchronize AC generators, several important factors must be checked:

- The phase rotation of both generator systems must be the same. Check this with lights as described later or use a phase rotation meter to determine ABC or ACB rotation.

- The AC voltages of both generators should be equal. In practice, the voltage of the incoming generator is usually 1 to 2 volts higher than that of the other operating generator.

- The frequencies of the incoming generators must match when synchronized. In practice, the frequency of the incoming generator is 1 to 2 Hz higher than that of the on-line generator. This can be observed with lights or by using a synchroscope.

The speed and output voltage of the incoming generator are slightly higher to prevent it from becoming a load to the system when it is connected.

Two methods of synchronization using lights are described next.

Three Dark Method

This section describes the method of synchronizing two alternators using the *three dark method*. Figure 14-2 illustrates a circuit used to parallel two three-phase alternators. Alternator G_2 is connected to the load circuit. Alternator G_1 is to be paralleled with alternator G_2. Three lamps rated at double the output voltage to the load are connected between alternator G_1 and the load circuit as shown. When both machines are operating,

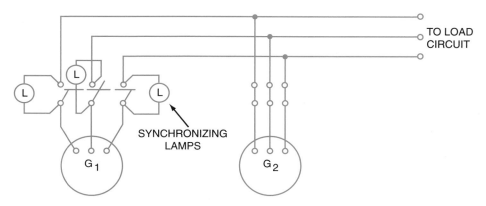

Figure 14-2 Synchronization of alternators.

one of two effects will be observed:

- The three lamps will light and go out in unison at a rate that depends on the difference in frequency between the two alternators.

- The three lamps will light and go out at a rate that depends on the difference in frequency between the two machines, but not in unison. In this case, the machines are not connected in the proper phase sequence and are said to be out of phase. To correct this, you need to interchange any two leads to alternator G_1. The machines are not paralleled until all lamps light and go out in unison. The lamp method is shown for greater simplicity of operation.

By making slight adjustments in the speed of alternator G_1, the frequency of the machines can be equalized so that the synchronizing lamps will light and go out at the lowest possible rate. When the three lamps are out, the instantaneous electrical polarity of the three leads from G_1 is the same as that of G_2. At this instant, the voltage of G_1 is equal to and in phase with that of G_2. Now the paralleling switch can be closed so that both alternators supply power to the load. The two alternators are in synchronism, according to the three dark method.

The three dark method has certain disadvantages and is seldom used. A large voltage may be present across an incandescent lamp even though it is dark (burned out). As a result, it is possible to close the paralleling connection while there is still a large voltage and phase difference between the machines. For small capacity machines operating at low speed, the phase difference may not affect the operation of the machines. However, when large capacity units having low armature reactance operate at high speed, a considerable amount of damage may result if there is a large phase difference when you attempt to parallel the units.

Two Bright, One Dark Method

Another method of synchronizing alternators is the two *bright, one dark method*. In this method, any two connections from the synchronizing lamps are crossed after the

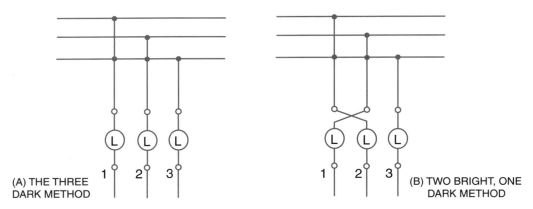

(A) THE THREE
DARK METHOD

(B) TWO BRIGHT, ONE
DARK METHOD

Figure 14-3 Methods of synchronizing alternators.

alternators are connected and tested for the proper phase rotation. (The alternators are tested by the three dark method.) Figure 14-3(A) shows the connections for establishing the proper phase rotation by the three dark method. Figure 14-3(B) shows the lamp connections required to synchronize the alternator by the two bright, one dark method.

When the alternators are synchronized, lamps 1 and 2 are bright and lamp 3 is dark. Because two of the lamps are becoming brighter as one is dimming, it is easier to determine the moment when the paralleling switch can be closed. Furthermore, by observing the sequence of lamp brightness, you can tell whether the speed of the alternator being synchronized is too slow or too fast.

Synchroscope

A synchroscope is recommended for synchronizing two alternators because it accurately shows the exact instant of synchronism (Figure 14-4). The pointer rotates clockwise when an incoming alternator is running fast and counterclockwise when an incoming alternator is running slow. When the pointer is stationary, pointing upward, the alternators are synchronized. Because the synchroscope is connected across one phase only, it cannot be used safely until the alternators have been tested and connected together for the proper phase rotation. Synchronizing lamps or other means must be used to determine the phase rotation. In commercial applications, the alternator connections to a three-phase bus through a paralleling switch are permanent. This means that continuous tests for phase rotation are not necessary. As a result, a synchroscope is the only instrument required to detect when the alternators are synchronized and to parallel them. However, a set of lights is often used as a double-check system.

Prime Movers

In industrial applications, alternators are driven by various types of prime movers such as steam turbines, water turbines, and internal combustion engines. For applications

(A)

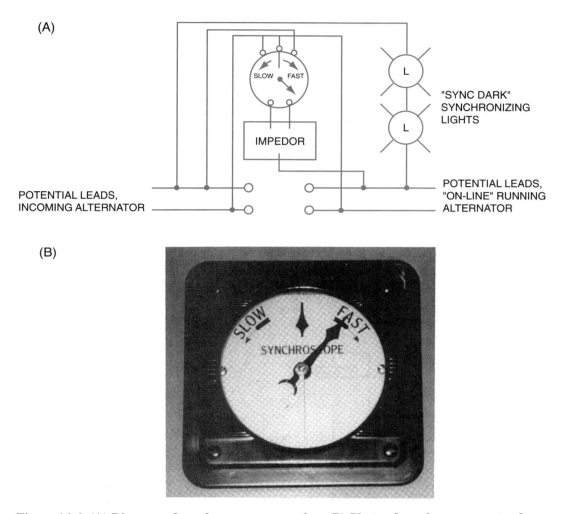

(B)

Figure 14-4 (A) Diagram of synchroscope connection. (B) Photo of synchroscope meter face.

on ships, alternators often are driven by DC motors. Regardless of how alternators are driven, speed variation is a factor in paralleling the machines. Thus, the electrician should have a basic knowledge of speed governors and other speed regulating devices. This text, however, does not detail the operation of these mechanical devices.

PARALLELING ALTERNATORS

Because apprentices are likely to be required to parallel alternators driven by DC motors sometime in their instruction, the following steps outline the procedure for paralleling these machines. Figure 14-5 illustrates a typical circuit for paralleling two three-phase alternators.

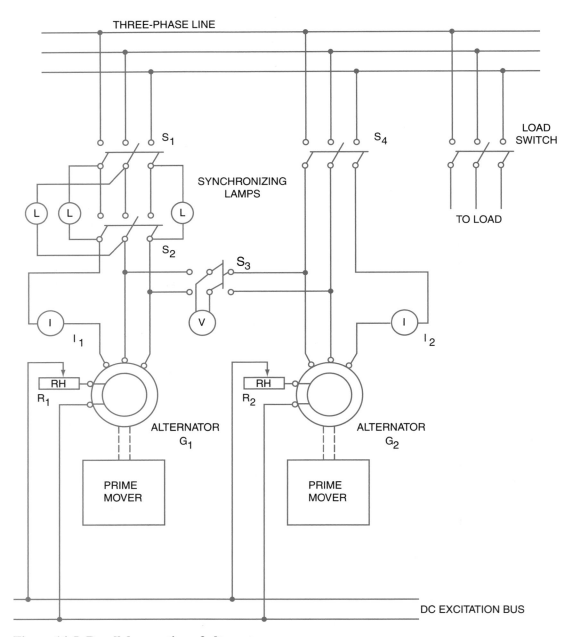

Figure 14-5 Parallel operation of alternators.

Procedure

1. Set the field rheostat (R_2) of alternator G_2 to the maximum resistance position.
2. Knowing the number of field poles in alternator G_2, determine the speed required to generate the desired frequency.

3. Energize the prime mover to bring alternator G_2 up to the required speed.

4. Set Switch S_3 to read the AC voltage across one phase of G_2. Adjust field rheostat R_2 until the output voltage is equal to the rated voltage of the load circuit.

5. Close the load switch and switch S_4 to feed the load circuit. Readjust the speed of the prime mover to maintain the predetermined speed required for the desired frequency.

6. Readjust R_2 to obtain the rated AC voltage of the load circuit.

7. Energize the prime mover to drive the second alternator, G_1. Adjust the speed of the alternator to the approximate value required to match the frequencies of the alternators.

8. Set switch S_3 to measure the AC voltage across one phase of G_1. Adjust field rheostat R_1 until the AC voltage is equal at either position of switch S_3. The voltage output of both alternators is now equal.

9. (Phase Rotation.) With paralleling switch S_2 open, close switch S_1. The three sets of lamps across the terminals of the open switch will respond in one of two ways:

   a. The three lamps will brighten and then dim in unison.

   b. Two lamps will brighten in unison as the remaining lamp dims. Then the two bright lamps will dim as the dark lamp brightens.

10. If the lamps respond as in 9(a), the alternators are connected for the proper phase rotation. The operator then may proceed to step 12 in synchronizing the alternators.

11. If the lamps respond as in 9(b), the alternators are not in the proper phase rotation. To correct the condition, interchange any two alternator leads at the terminals of switch S_2. All three lamps should dim together and brighten together. No attempt to parallel the alternators should be made until the lamps respond in this manner.

12. The three lamp sets will flicker (dim and brighten) at a rate equal to the frequency difference between the two alternators. Adjust the speed control of prime mover M_1 to make the lamps flicker at the lowest possible rate.

13. Interchange two lamp set leads (not alternator leads) at the terminals of switch S_2 so that the alternators can be synchronized using the two bright, one dark method.

14. Again adjust the field rheostat of alternator G_1 until both alternators have the same output voltage as measured at either position of the voltmeter switch S_3.

15. With one hand on switch S_2, watch the lamps. Close the switch at the exact instant that two lamps are at their brightest and the other lamp is out. This operation shunts out the synchronizing lamps and parallels the alternators.

16. Ammeters I_1 and I_2 indicate the amount of load current carried by each alternator. If the load circuit has a unity power factor, then the sum of the ammeter readings should equal the reading of the ammeter in the load circuit.

17. Note that a change in the field excitation of either alternator does not appreciably change the amount of current supplied to the system. Such a change in field excitation does, however, affect the power factor of the specific alternator. The field rheostat of each machine should be adjusted to the highest power factor as indicated by the lowest value of current from the individual machine. Increasing or decreasing the mechanical power to either alternator will increase or decrease the load current of that machine. As a result, the division of the load between the alternators can be changed by slight changes in the alternator speed.

Speed versus Load Characteristics

Two alternators operating in parallel must have the same frequency and the same terminal voltage. In addition, the prime movers of the parallel alternators must have similar drooping speed load characteristics. For steam-, diesel-, water-, or gas-driven prime movers, the speed load characteristic depends on adjustments of a mechanical speed control governor. These adjustments determine the division of load for two alternators operating in parallel. For this reason, the kilowatt load delivered by two alternators in parallel cannot be divided in any desired proportion by varying the DC field excitation of either machine.

Two alternators properly connected in parallel will operate in stable equilibrium. If one alternator attempts to pull out of synchronism, a current is created that circulates between both alternators. This current increases the speed of the lagging machine and retards the leading machine thus preventing the machines from pulling out of synchronism.

REVERSE POWER

If, for any reason, one machine is allowed to slow to a point where the other machine is taking all the electrical load, the zero load generator then goes to a negative value or "reverse power." This generator has now become a motor. This situation is of particular concern where the machine's protective scheme has not been designed to operate properly in the motoring situation. In such conditions, reverse-current relays are usually employed to trip the generator on detection of reverse power flow.

Results of Motorization of a Generator

If a generator loses prime mover power, it acts as a motor with a DC field on the rotor. The DC field will cause the rotor to try and follow the AC field in the same direction as before. If the mechanical drag on the rotor is heavy, it will fall behind and "slip poles," inducing a large voltage into the rotor; this can cause insulation breakdown of the windings, flashover at the brushes, and violent shaking of the generator mountings.

If a generator loses DC excitation to the rotor it will not generate, but the prime mover power will still turn the rotor. Now the generator acts as a motor running at no load.

SUMMARY

In many cases, the parallel operation of alternators is essential to provide needed power and to maintain electrical power during peak loads or when removing an alternator from service for maintenance. The requirements for paralleling are (1) the phase rotation of the generated voltage must be the same, (2) the voltage at the paralleling point must be the same, and (3) the frequencies of the generators must match. Lamps and synchroscopes are often used to aid in the paralleling procedure. If a generator loses output power, it must be removed from the electrical power system or serious consequences could result.

ACHIEVEMENT REVIEW

A.   Select the correct answer for each of the following statements and place the corresponding letter in the space provided.

1. Two alternators are paralleled   _____
    a. so that one is not overworked.
    b. because of a rising load demand.
    c. to ease the workload.
    d. because of the declining load demand.

2. To parallel alternators, it is necessary to match   _____
    a. voltages.
    b. frequencies.
    c. voltages and frequencies.
    d. voltages, frequencies, and instantaneous polarities.

3. The output voltage of an alternator is controlled by   _____
    a. adjusting the prime mover.
    b. adjusting the direct current of the field circuit.
    c. synchronizing lamps.
    d. a synchroscope.

4. Alternators should not be paralleled unless the synchronizing   _____
    lamps are lighting and dimming
    a. in rotation.
    b. in reverse rotation.
    c. in unison.
    d. alternately.

5. Three lights flashing rapidly in unison while paralleling   _____
    alternators means that
    a. the machines are not polarized.
    b. the phase sequences are wrong.
    c. the paralleling switch should be closed.
    d. the frequencies differ by a large amount.

6. The three dark method of synchronizing alternators has _____
   the disadvantage that
   a. the lamps may burn out.
   b. an undetected voltage may be present at the lamps.
   c. the light is more difficult to see.
   d. an undetected current may be present through the lamps.

7. The most reliable method of synchronizing alternators is to use _____
   a. a synchroscope.
   b. the three dark method.
   c. the three light method.
   d. the two bright, one dark method.

8. If a synchroscope is rotating clockwise, the _____
   a. alternators are ready to parallel.
   b. alternator being synchronized is too slow.
   c. alternator being synchronized is too fast.
   d. machines have not been polarized.

9. When the pointer of a synchroscope is stationary and points _____
   upward during the paralleling operation, the
   a. alternators are in synchronism.
   b. alternators are not in synchronism.
   c. incoming alternator frequency is too slow.
   d. incoming alternator frequency is too fast.

10. The division of load between alternators operating in parallel _____
    is accomplished by changing the
    a. field excitation.
    b. speed of the prime movers.
    c. power factor of the load.
    d. machine characteristics.

B.   Insert the word or phrase to complete each of the following statements.

11. To operate satisfactorily in parallel, two alternators must have the same _____,
    the same frequency, and the same_____.

12. Two alternators are to be connected in parallel. The best instrument to use for syn-
    chronizing them is a(n) _____.

13. An alternator is connected to a live three-phase bus. Using the three dark method,
    a lamp is connected in series with each lead. The lamps brighten and dim in unison.
    This proves that the alternators have the proper _____ rotation.

14. In question 3, the switch shorting the three series lamps should be closed at the
    instant the lamps are _____.

15. Two 208-volt alternators are to be paralleled. The synchronizing lamps should be rated at _____.

16. The output voltage of alternators operating in parallel is equalized by adjusting their _____.

17. The load on an alternator operating in parallel with another alternator may be increased by decreasing the spring tension of its speed _____.

18. The division of load between two alternators operating in parallel can be changed by adjusting the _____.

19. Two alternators, A and B, are being synchronized for parallel operation. Alternator A is operating at a frequency of 60 hertz. The synchronizing lamps are flickering twice a second. The frequency of alternator B is_____ hertz or _____ hertz.

20. Synchronizing lamps and a synchroscope are being used to parallel two alternators. Just before the moment the alternators are paralleled, there is no visible light from the lamps but the synchroscope is rotating slowly. In this case, the _____ method should be used to indicate when the paralleling switches should be thrown because _____.

U•N•I•T
15
WIRING FOR ALTERNATORS

OBJECTIVES

After studying this unit, the student should be able to

- describe the connections for and the resulting operation of the direct-current field excitation circuit for an alternator.
- describe the connections for and the resulting operation of the alternator output circuit for an alternator.
- describe the connections for and the resulting operation of the instrument circuits for an alternator.

This unit presents the control panel and equipment for a three-phase, 2,400-volt alternator. The circuits and connections covered in detail are the direct-current field circuit and all control equipment; the alternating-current, three-phase output circuit with associated switchgear; and the connections for the instruments and instrument transformers used in a common installation.

DIRECT-CURRENT CIRCUIT FOR FIELD EXCITATION

When alternators are connected so that they are separately excited, a DC bus must be used to deliver the DC power for field excitation to multiple generators. This system is used so that if one alternator fails, other alternators may still use the DC bus for excitation. In this case, the DC bus is connected to each of the alternators supplying power to the system. Each generator control panel has individual controls to monitor and adjust the DC power taken from the bus and delivered to the alternator's field. In this way each alternator's output voltage can be controlled and the power factor of the output power can be adjusted by the field control. In some generation systems, switches open and close the DC power to the field. If there are manual switches, the switch must contain some method of discharging the DC field as it is opened. If there is no system for discharging the field, the DC field that collapses in the field coil will generate very high voltages in the field coil (inductive kick concept) and the field windings may be destroyed. In some generator systems, a large diode system is installed to prevent field discharge damage. Under a normal connection, the diode is reverse biased and does not conduct. In the event of the magnetic field collapse, the diode becomes forward biased and conducts current back to the winding, reduces the inductive kick effect, and prevents field breakdown.

Figure 15-1 illustrates the connections required for the separately excited field circuit of an alternator. Note that when the field discharge switch is open, the auxiliary blade closes to complete a path through the field discharge resistor. Thus, any inductive voltage in the alternator field is discharged through the field discharge resistor to prevent damage. *The field rheostat is connected so that it is not in the discharge circuit.*

Figure 15-1
Separately excited
circuit for field
connections of
an alternator.

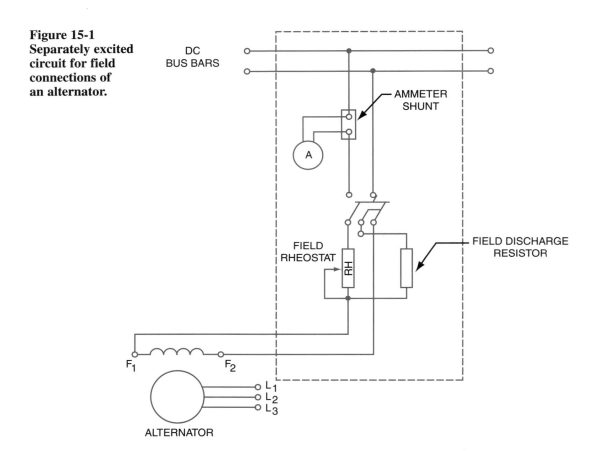

ALTERNATOR OUTPUT CIRCUIT

The alternator in the installation described in this unit is rated at 2,400 volts, three phase. The three-phase, 2,400-volt output of the alternator is fed to the switchboard through a three-wire, high-voltage lead cable in galvanized rigid conduit. The three conductors are fed through an oil-type circuit breaker, current transformers, and disconnect switches to the three-phase bus bars. An oil-type circuit breaker (switch) is used because of the relatively high voltage of the alternator. As the contacts of this switch open, any arc is immediately quenched in insulating oil.

Figure 15-2 illustrates an electrically operated oil switch (circuit breaker). Note that each of the three sets of contactors is mounted in a separate cell or tank that is filled

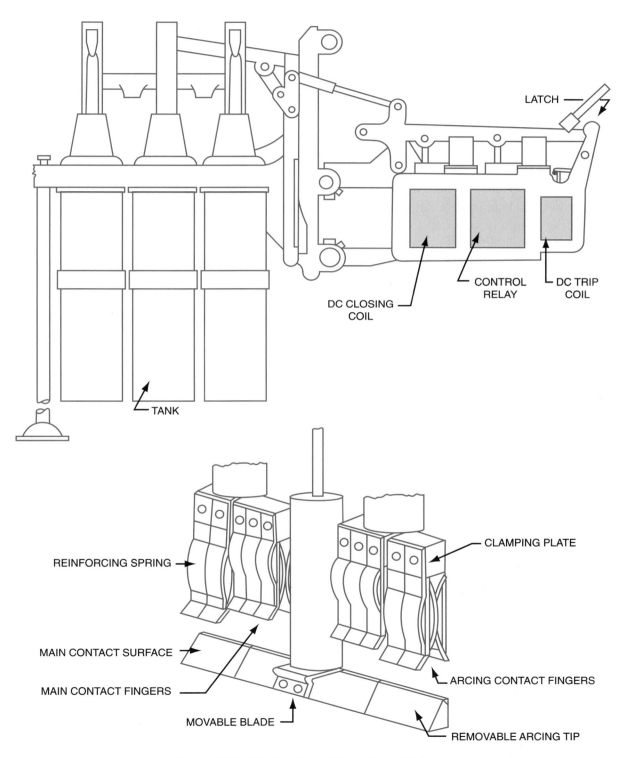

Figure 15-2 Details of an oil-type circuit breaker.

with an insulating oil. The three sets of contactors thus open and close in oil. The figure also shows a contactor assembly for one pole of a three-pole oil switch. Note the closing coil and the trip coils. The closing coil is relatively large and has a very fast positive action; the trip coil is smaller in size. The trip coil actuates a trip latch that causes the oil switch contactors to open.

The control circuit for the oil switch in most alternator installations is connected to a DC source such as a bank of batteries. If there is a complete failure of the AC power, the oil switch can still be operated from the DC source, as is true of other emergency circuits.

A small switch handle located on the switchboard is used to actuate the control circuit. Two indicating lamps also are mounted on the switchboard. One of the indicating lamps is green and is on when the oil switch is open. The second indicating lamp is red and is on when the oil switch is closed. The red lamp normally is located directly above the control switch handle, and the green lamp is located below the switch handle.

Figure 15-3 shows the schematic connection diagram of the control circuit for the oil switch. When the oil switch is in the open or off position, the green pilot lamp is on. Note that there is a path from the positive side of the line through the current-limiting resistor, through the green indicating lamp, and through the normally closed M contacts to the negative side of the line.

When the on (start) button is pressed, a circuit is established from the positive side of the line to the control relay and then to the negative side of the line. The control relay is energized and closes its contacts to establish a path through the DC closing coil. The three main sets of oil switch contacts also change at this time. When the main closing relay is energized, the normally closed M contacts open. In addition, the green pilot lamp circuit opens and the two normally open M contacts close. The red indicating lamp is now on. When the on button is released, the oil switch remains in the on position due to the fact that it is secured by a mechanical latch mechanism.

When the off button is pressed, the DC trip coil is energized to trip the latch mechanism. The oil switch contacts thus open to the off position. As a result, the red indicating lamp goes out and the green indicating lamp lights.

The control handle and indicating lamps for an oil switch generally are mounted on the switchboard. The oil switch itself, however, is usually located separately from the switchboard room.

Current transformers are used to step down the current in the output leads of the alternator to a value that can be used in instrument circuits. Step-down current transformers also insulate the low-voltage instrument circuit from the high-current primary circuit. The secondary current rating of a current transformer is 5 amperes (see Unit 23). The current rating of the primary winding of the transformer must be high enough to handle the maximum current delivered by the alternator.

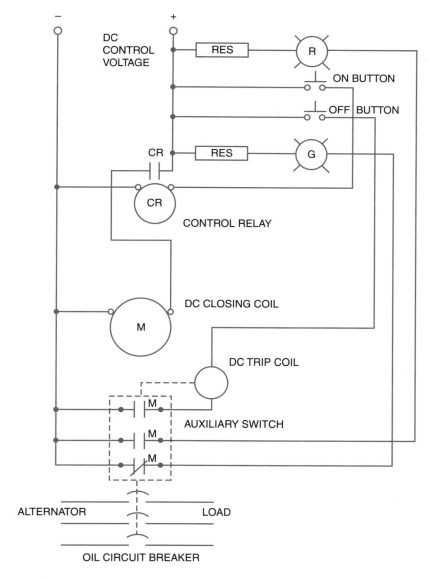

Figure 15-3 Elementary control circuit for an oil circuit breaker.

The alternator output leads feed from the current transformers to disconnect switches and then to the three-phase bus bars. A disconnect switch is a form of knife switch that is opened with a switch stick while exposed to air. The disconnect switches are operated *only* after the alternator oil switch is opened. The operator must wear rubber gloves when using an approved switch stick to open the disconnect switches. *Never open disconnect switches under load*—this is the purpose of the oil switch. It is designed to interrupt the arc without damage to the switch blade.

In most alternator installations, the three-phase bus bars are energized constantly. Because the disconnect switches disconnect the oil switch and the alternator from the bus bars, the alternator can be shut down and the disconnect switches opened to permit maintenance work on the oil switch under safe conditions. When the alternator requires maintenance or repair work, the disconnect switches are pulled to the off position even though the oil switch is open. The reason for this precaution is that the insulating oil in the oil switch may have become carbonized. The carbonized oil can act as a partial conductor resulting in a feedback from the live 2,400-volt bus bars through the oil switch and carbonized oil to the alternator terminals. Remember that the disconnect switches and the oil switch must be *open* when any maintenance or repair work is to be done on AC generators. The generators should also be shut down.

Figure 15-4 is a wiring diagram of typical alternator connections to the three-phase bus bars.

The three bus bars for the AC output of the alternator are mounted on insulators because the bus bars have a potential difference of 2,400 volts between them. It is important that the proper air gap be maintained between the three bus bars and that adequate clearance be provided between the bus bars and the ceiling and side walls of the room. Barriers shall be placed in all service switchboards to isolate the service bus bars and terminals from the remainder of the switchboard.

Large generators are constructed in two styles. One style uses a separate exciter DC generator and feeds the DC excitation field into the alternator rotor through brushes and slip rings. Because the field current and voltage are relatively low compared to the output of the alternator, brushes and slip rings work well. The other style of large generation equipment uses a brushless exciter style to supply DC to the rotor. Either method is effective and accomplishes the same task—to provide a DC field to the rotating field of the AC generator.

To adjust the field and provide the desired output voltage, the output voltage levels must be monitored. In the brush-type rotor connection, the AC is monitored at the output and a DC field of a small DC separately excited generator is controlled. As the output voltage drops, the DC field is increased. This small DC generator, called an *amplydine,* supplies the DC field to a larger DC exciter generator. The second generator then supplies DC to the alternator's field. This process allows for stages of amplification of the DC field. A small change in output AC affects the DC field to the amplydine that feeds the second stage of amplification for the DC to the alternator field. A small control voltage at the amplydine level is used to control the large DC to the rotor of the generator.

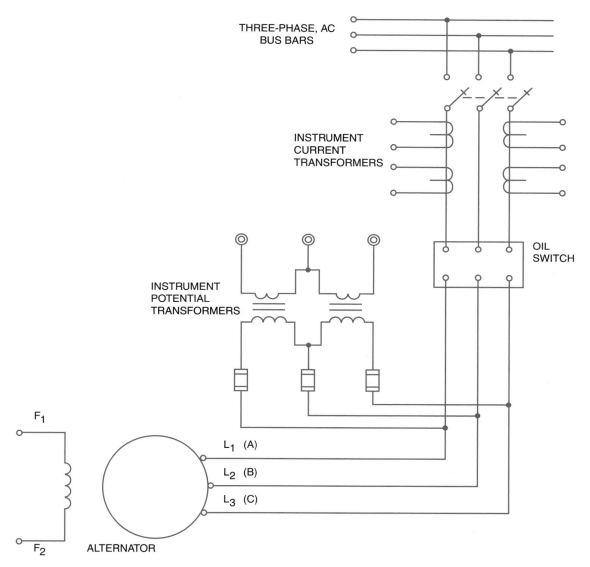

Figure 15-4 Wiring diagram of a three-phase alternator circuit.

Brushless exciters are discussed in Unit 11. The concept is to use a small amount of controllable DC, and then amplify it and feed it to the alternator field. This process uses semiconductors to change induced AC into DC on the rotor. Figure 15-5 shows a block diagram of the two styles of field control.

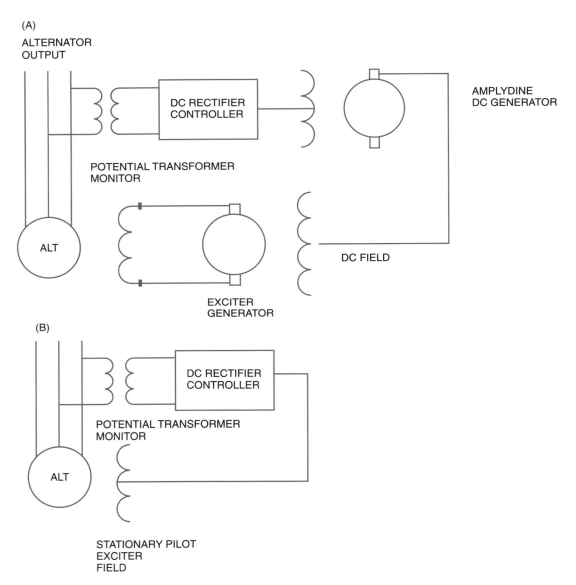

Figure 15-5 (A) Brush type of alternator using amplydine system. (B) DC is produced on the rotor in the brushless exciter by mounted rectifiers.

ALTERNATOR COOLING

The electrical current flowing in the alternators through conductors produces heat in the coils because of $I^2 R$ losses. This heat must be removed from the alternator to maintain operations. For most small alternators, the windings are air cooled by a fan on the rotor that pushes air through slots in the armature winding core. If the alternators are indoors, means must be provided to vent the heat as the generators run. In large utility generators, air cooling is not enough. In these circumstances, hydrogen is used for cooling the alternator core. Using hydrogen for cooling has the advantage of being less dense than air so that it reduces the windage losses in the alternator cooling process. Hydrogen also absorbs and removes more heat than air, which provides more efficient cooling. This hydrogen system is normally passed around the output voltage bus as well to provide cooling for the output conductors of the alternator.

INSTRUMENT CIRCUITS

The voltage to the potential coils of instruments mounted on the switchboard should not exceed 120 to 125 volts. The voltage coils of wattmeters, watt-hour meters, and voltmeters usually are designed for a maximum voltage of 150 volts. Because the three-phase output of the alternator is 2,400 volts, two instrument potential transformers connected in open delta are required to step down the voltage to 120 volts, three phase (see Unit 23). The potential transformers are small in size because the load on the low-voltage secondary is very small. Each potential transformer is rated at 100 to 200 volt-amperes (VA). For the installation shown in Figure 15-6, the load on the secondary of the transformer consists of the potential coils of the kilowatt meter and the voltmeter. The instrument potential transformers are rated at 2,400 volts on the high-voltage side and 120 volts on the low-voltage side. The low voltage at the instruments allows maintenance electricians to work more safely when making adjustments and repairs to the instruments.

The current coils of the measuring instruments mounted on switchboards are rated at a maximum current capacity of 5 amperes. In Figure 15-6, each of the two current coils of the three-phase kilowatt meter is connected in series with the proper current transformer.

It is unsafe to open the secondary circuit of a current transformer when there is a current flow in the primary circuit. (See Unit 23.)

Figure 15-6 shows a wiring diagram for most of the instruments and instrument transformers described. The current in the secondary of current transformer circuits is never in excess of 5 amperes. Therefore, either 14 or 12 AWG wire is used on the rear of the switchboard.

For most permanent switchboard installations, the scale readings on the instruments are graduated to include the voltage and current transformer multipliers. This means that any error made by the switchboard operator in applying instrument multipliers is automatically eliminated.

Two instruments not shown in the wiring diagram in Figure 15-6 are the voltmeter and the synchroscope. In typical installations, several alternators may be operating in parallel. Each alternator has a separate panel, and these panels are mounted next to one another to make up a complete switchboard. One voltmeter and one synchroscope are then mounted on a movable panel located at the end of the switchboard. The position of this panel can be adjusted so that the voltmeter and synchroscope are visible from any

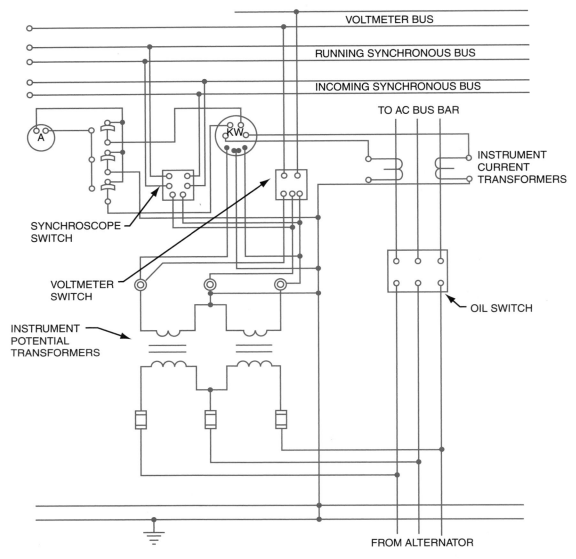

VOLTMETER BUS

RUNNING SYNCHRONOUS BUS

INCOMING SYNCHRONOUS BUS

TO AC BUS BAR

INSTRUMENT
CURRENT
TRANSFORMERS

SYNCHROSCOPE
SWITCH

VOLTMETER
SWITCH

INSTRUMENT
POTENTIAL
TRANSFORMERS

OIL SWITCH

FROM ALTERNATOR

Figure 15-6 A wiring diagram for instruments and potential transformers.

one of the generator control panels. A voltmeter switch located on each generator panel gives the operator a means of connecting the voltmeter to measure the voltage output of any alternator. In addition, special synchronizing switches permit the use of one synchroscope to synchronize any one of several alternators to the three-phase system.

Figure 15-7 shows the circuit connections for the voltmeter and synchroscope. Figure 15-6 indicates that the voltmeter switch has three positions. The voltmeter can be connected across any one of the three voltages of an alternator. If the voltage of a second

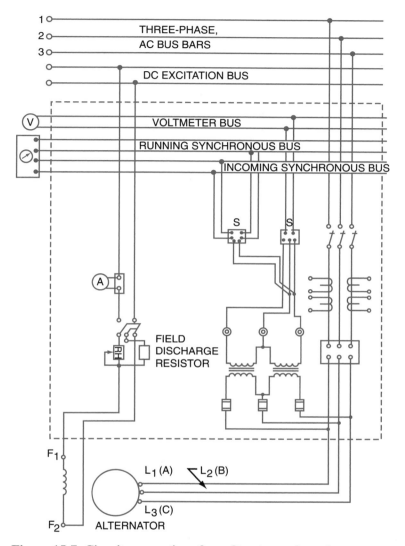

Figure 15-7 Circuit connections for voltmeter and synchroscope.

alternator must be measured, the voltmeter switch is turned to the off position. The switch handle or key is then removed and inserted in the voltmeter switch of the second AC generator. Again, the switch may be turned to any one of the three voltage positions. Thus, one voltmeter can be used to measure the three voltages of each of several AC generators controlled through the switchboard.

A synchroscope switch is mounted on each alternator panel. When the switch handle is turned to the incoming position, the synchroscope is connected to the secondary voltage of one phase of an alternator being synchronized with the AC system. The synchroscope switch of a second alternator, which is already paralleled with the three-phase system, is connected to the run position. Thus, one coil winding of the synchroscope is energized from the running bus bars. The other winding of the synchroscope is energized from the incoming bus bars. With these connections, the synchroscope will indicate the extent that the incoming machine is out of phase. When the incoming alternator is in phase with the three-phase system and the alternator voltage is equal to that of the bus bars, the control switch can be turned to the on position. As a result, the oil switch contactors close and the alternator is paralleled with the bus bars. The oil circuit breaker is used to connect and disconnect the alternator when it is running under load. This ensures safe operation and prolongs switch contact life.

Many of the meters on generators are now in digital format. This allows the user to more precisely determine the exact currents and voltages being produced. The use of these meters is still supplemented by the synchroscope to verify the exact point of synchronism if manual paralleling of the alternators is needed.

SUMMARY

Connections for the alternator include the input power in the form of DC field excitation and the output power in the form of AC-generated power. DC can be supplied through a DC exciter bus. The connection to an individual generator's exciter field would then be through a field switch. The field switch must operate to supply DC to the magnetic field and also provide for the disconnection and magnetic field discharge. The output power of a generator with higher generated voltages may be through a switch designed to quench the arc when disconnecting. One such switch is the oil-type circuit breaker. These switches have arc control systems designed for the rated voltage and current. The load-break switch is often held by a latching control circuit so that the breaker will stay closed without power consumption. Instrument circuits are used to monitor the electrical functions of the alternators and provide feedback for voltage regulation and current monitoring.

ACHIEVEMENT REVIEW

1. What is the purpose of disconnect switches in an AC generator installation? _____

   _____

   _____

   _____

   _____

   _____

2. Why is direct current used on the control circuits of oil switches used in alternator installations? _____

_____

_____

_____

_____

_____

3. Why is an oil switch normally used to interrupt the power output of an alternator?

_____

_____

_____

_____

4. Why are instrument transformers used for the instrument circuits of AC generator installations?_____

_____

_____

_____

_____

_____

16

SUMMARY REVIEW
OF UNITS 11–15

OBJECTIVE

- To provide the student with an opportunity to evaluate the information and understanding acquired in the study of the previous five units.

A. Insert the word or phrase to complete each of the following statements.

1. The main three-phase leads from a high-voltage alternator usually feed the main bus bars through a switch, the contacts of which are covered by _____.

2. To minimize the danger to personnel working on the maintenance of high-voltage, three-phase alternators, a _____ switch is used in the main three-phase output leads.

3. If the DC supply fails that is used to control the main switch of a three-phase, high-voltage alternator, a separate DC source consisting of _____ is used.

4. An indicating lamp is used to indicate that the main switch is closed on a three-phase, high-voltage alternator. This lamp is colored _____.

5. A _____ indicating lamp is used to indicate that the main line switch is open.

6. Current is measured in the three-phase leads of a high-voltage alternator by ammeters connected in the output leads of the alternator through the _____.

7. The disconnect switch in the main line of an alternator is opened and closed by an operator using rubber gloves and a _____.

8. A voltage feedback from the main bus to the output terminals of the alternator can occur through an open oil switch as the oil becomes _____.

9. Voltage measurements are made on high-voltage alternators with voltmeters connected to the line through _____.

10. The regulation of an alternator is influenced by the impedance of its windings and the _____ of the load circuit.

11. The speed of the prime mover driving an alternator determines the _____ and _____ of the output.

135

12. The normal voltage regulation of an alternator is least affected by a load with a slightly _____ power factor.

13. The output voltage of alternators is maintained through the use of voltage _____.

14. When paralleling two alternators, the procedure used to bring both machines to the same exact phase relationship is called _____.

15. The paralleling of alternators without a synchroscope is best accomplished with synchronizing lamps using the _____ method.

16. The voltage output of an alternator is controlled by adjusting the _____ circuit resistance.

17. In a revolving-field alternator, slip rings are used to conduct current to the _____ circuit.

18. A DC generator mounted on the same shaft as the alternator is referred to as the _____.

19. At a fixed speed of rotation, the frequency of the output voltage depends on the number of _____.

20. Alternator field windings are marked with the letters _____ and _____.

21. The extent to which voltage output decreases with increases in load current is referred to as voltage _____.

22. The three-phase windings and the laminated core of a three-phase alternator of the rotating-field type are known as the _____.

23. An alternator with four field poles is to generate power at 60 Hz. For this frequency the speed must be _____ RPM.

24. An increase in the field current of an alternator increases its output voltage to an extent determined by field _____.

25. The regulation of an alternator is poorest when the load circuit has a low, _____ power factor.

26. A conventional UPS system used to supply power during utility power interruption or fluctuations consists of a battery system, a charging system, and an _____.

27. Wind farms are not designed as base load generators but as _____ generation.

28. When connecting fuel cells for supplemental power, Code® Article _____ states that there must be a sign warning of alternate power sources available on the site.

29. Flywheel UPS systems are designed for use with a _____ for long duration interruptions.

B. Select the correct answer for each of the following statements and place the corresponding letter in the space provided.

30. When *removing* the load from an alternator, the   _____
    a. oil switch should be opened.
    b. disconnect switch should be opened first.
    c. machine should be slowed.
    d. machine should be stopped.

31. Current to the ammeters on an alternator installation   _____
    switchboard is never in excess of
    a. 100 amperes.
    b. 2,400 volts.
    c. 50 amperes.
    d. 5 amperes.

32. A main disconnect switch is used to   _____
    a. remove the load from the alternator.
    b. disconnect the oil switch and alternator from the
       energized bus bar.
    c. energize the bus bar.
    d. energize the oil switch and alternator.

33. Alternator installation switchboard voltmeters are   _____
    connected to
    a. potential transformers.
    b. current transformers.
    c. the hot bus bar.
    d. the oil switch.

34. An oil switch is used to   _____
    a. remove the disconnect switch from the line.
    b. energize the alternator.
    c. interrupt high voltages and currents.
    d. lubricate the disconnect switch.

35. When using a synchroscope to parallel alternators, the   _____
    switches are closed when the indicator is
    a. revolving clockwise.
    b. revolving counterclockwise.
    c. pointing straight up.
    d. oscillating.

36. Adjusting the speed of the prime mover of an alternator   _____
    causes a change primarily in the
    a. voltage.
    b. frequency.
    c. phase polarity.
    d. phase poles.

37. The voltage output of an alternator should be increased   _____
    or decreased by
    a. adjusting the field rheostat.
    b. adjusting the speed.
    c. changing the number of poles.
    d. changing the capacities.

38. In an automatic transfer switch, the purpose of the   _____
    time delay relay is to
    a. allow the engine-driven generator to pick up speed.
    b. permit the load to increase.
    c. delay the normal power supply until it is firmly established.
    d. delay the emergency power supply until it is firmly established.

39. Microturbines use an electromechanical generation to produce power. This output
    is _____.
    a. exactly 60 Hz so no inverter is needed.
    b. variable frequency so is rectified.
    c. a high fixed frequency, which is chopped.
    d. DC, so is filtered and then inverted.

U • N • I • T
17

BASIC PRINCIPLES OF
TRANSFORMERS

OBJECTIVES

After studying this unit, the student should be able to

- explain how and why transformers are used for the transmission and distribution of electrical energy.
- describe the basic construction of a transformer.
- distinguish between the primary and secondary windings of a transformer.
- list, in order of sequence, the various steps in the operation of a step-up transformer.
- make use of appropriate information to calculate the voltage ratio, voltages, currents, and efficiency for step-up and step-down transformers.
- explain how the primary load changes with the secondary load.

It is neither efficient nor economically feasible to generate large quantities of *direct-current* electrical energy. The invention of the transformer was a milestone in the development of the electrical industry. The transformer increases or decreases the voltage of large quantities of *alternating-current* energy efficiently, safely, and conveniently. A large power distribution station is shown in Figure 17-1.

Figure 17-1 Substation with three oil-filled circuit breakers.

Large amounts of alternating-current energy may be generated at a convenient voltage using steam, nuclear, or water power. Transformers are used first to increase this energy to a high voltage for transmission over many miles of transmission wires, and then to decrease this voltage to values that are convenient and safe for use by the consumer.

ELEMENTS OF TRANSFORMERS

A transformer consists of two or more conductor windings placed on the same iron core magnetic path, as shown in Figure 17-2.

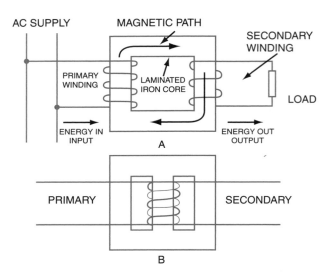

Figure 17-2 Parts of a transformer.

Laminated Core

The iron core of a transformer is made up of sheets of rolled iron. This iron is treated so that it has a high magnetic conducting quality (high permeability) throughout the length of the core. *Permeability* describes the ease with which a material will conduct magnetic lines of force. The iron also has a high ohmic resistance across the plates (through the thickness of the core). It is necessary to laminate the iron sheets (Figure 17-3) to reduce hysteresis and eddy currents that cause heating of the core.

Windings

A transformer has two windings: the primary winding and the secondary winding. The *primary winding* is the coil that receives the energy. It is formed, wound, and fitted over the iron core. The *secondary winding* is the coil that provides the energy at a transformed or changed (increased or decreased) voltage.

Transformers by definition are used to transfer energy from one AC system to another by electromagnetic means. They do not change the amount of power significantly; only minor wattage losses occur in the transformer. If the transformer increases

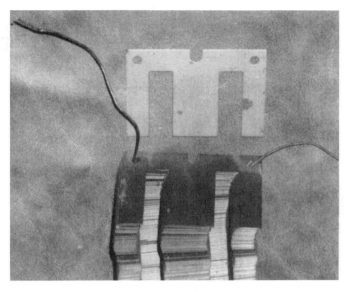

Figure 17-3 "E" lamination used in transformer core construction.

the voltage, it is called a *step-up* transformer. If it decreases the voltage, it is called a *step-down* transformer.

The secondary voltage is dependent on the following:

- voltage of the primary
- number of turns on the primary winding
- number of turns on the secondary winding

Certain core-type transformers have the primary and secondary wire coils wound on separate legs of the core [see Figure 17-2(A)]. The primary and secondary wire coils can also be wound on top of one another, as shown in Figure 17-2(B). Winding in this manner improves transformer efficiency and conserves energy. When stating the transformer ratio, the primary is the first factor of the ratio. This indicates which winding, high or low, is connected to the power source.

CONSTRUCTION OF TRANSFORMERS

As shown in Figure 17-4, three major types of construction for transformer cores are core, shell, and cross (or H).

Core Type

In a core transformer, the primary winding is on one leg of the transformer and the secondary winding is on the other leg. A more efficient type of core construction is the shell, in which the core is surrounded by a shell of iron [Figure 17-4(A) and (B)].

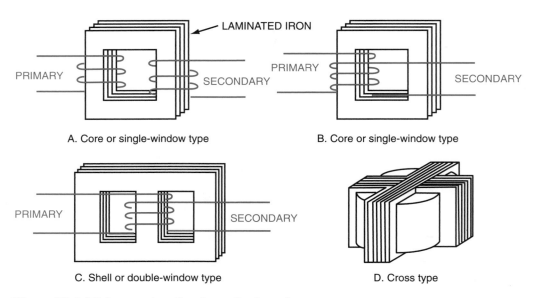

A. Core or single-window type

B. Core or single-window type

C. Shell or double-window type

D. Cross type

Figure 17-4 Major construction types for transformer cores.

Shell Type

The shell or double-window core transformer [Figure 17-4(C)] is probably used most frequently in electrical work. In terms of energy conservation, this transformer design operates at 98 percent or higher efficiency.

Cross or H Type

The cross or H core is also called the modified shell type [Figure 17-4(D)]. The coils are surrounded by four core legs. The cross type is really a combination of two shell cores set at right angles to each other. The windings are located over the center core, which is four times the area of each of the outside legs. This type of core is very compact and can be cooled easily. It is used for large power transformers where voltage drop and cost must be kept to a minimum. These units are usually immersed in oil for high insulation properties and effective cooling. Another method of cooling the transformers is by forced air. Transformers should never be immersed in water for cooling. Accidental flooding, such as in underground transformer vaults, should be pumped.

ELEMENTARY PRINCIPLES OF TRANSFORMER OPERATION

According to Lenz's law, a voltage is induced in a coil whenever the coil current is increased or decreased. This induced voltage is always in such a direction as to oppose the force producing it. Called *induction,* this action is illustrated by arranging two loops of wire, as shown in Figure 17-5.

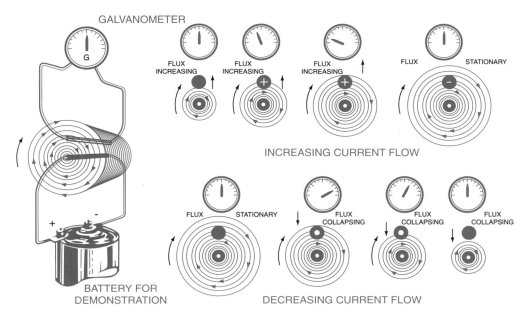

GALVANOMETER

FLUX INCREASING FLUX INCREASING FLUX INCREASING FLUX STATIONARY

INCREASING CURRENT FLOW

FLUX STATIONARY FLUX COLLAPSING FLUX COLLAPSING FLUX COLLAPSING

BATTERY FOR DEMONSTRATION   DECREASING CURRENT FLOW

Figure 17-5 Magnetic induction (electron flow).

Note in Figure 17-5 the progressive enlargement of the magnetic field about one side of each loop as the current builds up. The strength of the magnetic field increases as the electrical current through the conductor increases from the power source. Figure 17-5 also shows the field pattern during the period that the current decreases.

Figure 17-5 uses the left-hand rule for conductors. Grasp the conductor in your left hand with your thumb extended in the direction of the electron flow. Your fingers will indicate the direction of the magnetic flux. The flux expands outward from the conductor as the current flow increases, and contracts toward the conductor center as the current flow diminishes.

As the current builds up to its maximum value, the circular magnetic lines around the wire move outward from the wire. This outward movement of magnetic lines of force cuts across the conductor of the second loop. As a result, an EMF is induced and current circulates in the loop, as indicated on the galvanometer located above the conductor.

When the current reaches its steady state in the first circuit, the flux is stationary and no voltage is induced in the circuit. The galvanometer indicates zero current.

When the battery circuit is opened, current falls to zero and the flux collapses. The collapsing flux cuts through the second circuit and again induces an EMF. The second induced current has a direction opposite to that of the first induced current, as indicated by the galvanometer needle. The final stage shows a steady state with no field and no induced current. This action is automatic with AC applied.

The loops of wire may be replaced by two concentric coils (loops with many turns) to form a transformer. Figure 17-6 shows a transformer that has a primary winding, an iron core, and a secondary winding. When a changing or alternating current is delivered to the primary winding, the changing primary current produces a changing magnetic field in the iron core. This changing field cuts through the secondary coil and thus induces a voltage; the value depends on the number of conductors in the secondary coil cut by the magnetic lines. This is called mutual inductance. Commercial transformers generally have fixed cores that provide complete magnetic circuits for efficient operation where there is little flux leakage and high mutual induction.

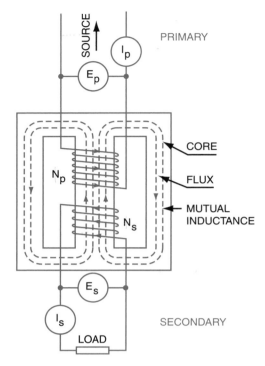

Figure 17-6 Single-phase transformer showing mutual inductance of two cells.

VOLTAGE RATIO

According to Lenz's law, 1 volt is induced when 100,000,000 magnetic lines of force are cut in 1 second. The primary winding of a transformer supplies the magnetic field for the core. The secondary winding, when placed directly over the same core, supplies the load with an induced voltage that is proportional to the number of conductors cut by the primary flux of the core.

The shell transformer type shown in Figure 17-6 is designed to reduce the voltage of the power supply. In Figure 17-6,

T_p N_p = number of turns in the primary winding

T_s N_s = number of turns in the secondary winding

I_p = current in the primary winding

I_s = current in the secondary winding

Assume that N_P = 100 turns

N_s = 50 turns

E_{supply} = 100 volts, 60 hertz

$$E_s = \frac{100 \times 50}{100}$$

$$E_s = \frac{5000}{100} = 50v$$

The alternating supply voltage (100 V) produces a current in the primary that magnetizes the core with an alternating flux. (According to Lenz's law, a counter EMF is induced in the primary winding. This counter EMF is called self-inductance and opposes the impressed voltage.) Because the secondary winding is on the same core as the primary winding, only 50 volts is induced in the secondary as only half as many conductors are cut by the magnetic field.

At no-load conditions, the following ratio is true:

$$\frac{N_p\ (100)}{N_s\ (50)} = \frac{E_p\ (100)}{E_s\ (50)}; \frac{2}{1} = \frac{2}{1}$$

The ratio of 2 to 1 indicates that the transformer is a step-down transformer that will reduce the voltage of the power supply. Transformers either step up or step down the supply voltage.

Refer to Figure 17-7 for the following example. The primary winding of a transformer has 100 turns, and the secondary has 400 turns. An EMF of 110 volts is applied to the primary. What is the voltage at the secondary, and what is the ratio of the transformer?

$$\frac{E_p}{E_s} = \frac{T_p}{T_s}$$

$$E_s = \frac{T_p}{T_s \times E_p}$$

$$\frac{E_p}{E_s} = \frac{N_p}{N_s}$$

$$\frac{110}{E_s} = \frac{100}{400}$$

$$100E_s = 44,000$$

$$E_s = \frac{44,000}{100} = 440 \text{ volts}$$

$$E_s = \frac{E_p \times T_s}{T_p}$$

$$E_s = \frac{110 \times 400}{100}$$

$$E_s = \frac{44000}{100}$$

$$E_s = 440 v$$

This transformer has a $\frac{440}{110} = \frac{4}{1}$ setup. The setup ratio is 1:4. $\frac{E_s}{E_p} = \frac{440}{110} = \frac{4}{1}$

CURRENT RATIO

The current ratio in a transformer is the inverse of the ratio for voltage transformation. The transformer does not create power and is not designed to consume power. The input power should be very close to the output power. Therefore, if the volt-amps input equals the volt-amps output and the voltage level is increased, the current level is decreased. The voltage ratio and the current ratio are inversely proportional.

If the load current of the transformer shown in Figure 17-7 is 12 amperes, the primary current must be such that the product of the number of turns and the value of the current (ampere-turns primary) equals the value of the ampere-turns secondary.

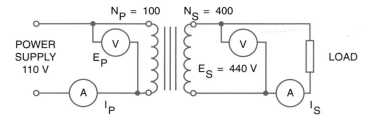

Figure 17-7 Elementary diagram of a transformer.

$$N_p I_p = N_s I_s \text{ or } \frac{N_p}{N_s} = \frac{I_s}{I_p}$$

$$\frac{100}{400} = \frac{12}{I_p}$$

$$100\, I_p = 4{,}800$$

$$I_p = 48 \text{ amperes}$$

Check of Solution for Current

$$N_p I_p = N_s I_s; \; 100 \times 48 = 400 \times 12; \; 4{,}800 = 4{,}800$$

The current ratio is an inverse ratio; that is, the greater the number of turns, the less the current for a given load. Practical estimates of primary or secondary currents are made by assuming that transformers are 100 percent efficient.

For example, assume that

$$\text{Watts input} = \text{Watts output}$$

or

$$\text{Primary watts} = \text{Secondary watts}$$

or

$$E_p \times I_p = E_s \times I_s$$

Therefore, for a 1,000-watt, 100/200-volt step-up transformer:

$$I_s = \frac{1{,}000 \text{ W}}{200 \text{ V}} = 5 \text{ amperes}$$

$$I_p = \frac{1{,}000 \text{ W}}{100 \text{ V}} = 10 \text{ amperes}$$

The greater the current, the larger size the wire leads are on the transformer. From this information, you can usually determine the high and low voltage sides.

Higher voltage = lower current, and therefore smaller wire size

Lower voltage = higher current, and therefore larger wire size

Example: A machine tool being relocated has a control transformer disconnected. The nameplate is illegible due to corrosion. The motor power circuit is 480 volts. The motor controller operates on 120 volts of control. Which is the primary and secondary of the control transformer? The higher voltage has the smaller wire size. Therefore, this is to be connected to the 480 volts.

The use of an ohmmeter can also tell you which winding has the greater resistance. By measuring each winding, you find that the greater the resistance, the higher the voltage connection because it has more turns of smaller wire. Remember, "primary" refers to the supply side of the transformer. "Secondary" refers to the load side (Figure 17-8).

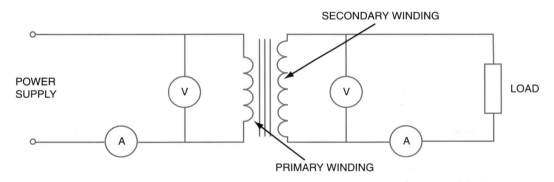

Figure 17-8 Schematic diagram of a step-up transformer.

SCHEMATIC DIAGRAM OR SYMBOL

A step-up transformer is usually shown in schematic form, as illustrated in Figure 17-8. The ratio of turns, primary to secondary, is not usually listed. This figure could be a step-up or step-down symbol representation.

PRIMARY LOADING WITH SECONDARY LOADING

The current in the secondary controls the current in the primary. When the secondary circuit is complete by placing a load across it, the secondary EMF causes a current to flow. This builds up a magnetic field in opposition to the primary field. This opposing, or demagnetizing, action reduces the effective field of the primary flux, which in turn reduces the primary CEMF, thereby permitting current to flow in the primary. The greater the current flow in the secondary, the greater the field produced by the secondary. This results in a reduced primary field; hence, a reduced primary CEMF is produced. This condition permits greater current flow in the primary. This entire process will repeat itself whenever there is any change in the value of the current in the secondary circuit. A transformer adjusts itself readily to any normal change in secondary load. However, if a direct short is placed across the secondary, the abnormally large amount of current flowing causes the primary current to rise in a like manner, resulting in damage to, or complete burn-out of, the transformer if it is not protected properly.

EFFICIENCY

The efficiency of all machinery is the ratio of the output to the input:

$$\text{Efficiency} = \frac{\text{output}}{\text{input}}$$

In general, transformer efficiency is about 97 percent. Only 3 percent of the total wattage at the secondary winding is lost through the transformation. The loss in wattage is due to *core losses* and *copper losses.*

The core loss is the result of hysteresis (magnetic friction) and eddy currents (induced currents) in the iron core.

The copper loss is power lost in the copper wire of the windings (I^2R). Therefore, taking these losses into consideration,

$$\% \text{ Efficiency} = \frac{\text{Watts output (secondary)}}{\text{Watts input (primary)}} \times 100$$

where Watts input = Watts output + losses.

SUMMARY

Transformers are very useful in delivering the exact voltage needed by customers. DC cannot be easily changed from one voltage level to another. There are no true DC transformers. AC can be increased or decreased easily through the electromagnetic coupling of the transformer coils. Transformers can be used to (1) step up the voltage; (2) step down the voltage; or (3) simply isolate the transformer primary system from the transformer secondary system.

ACHIEVEMENT REVIEW

A.   Select the correct answer for each of the following statements and place the corresponding letter in the space provided.

1.   When the primary winding has more turns than the   _____ C _____
secondary, the voltage in the secondary winding is
    a. increased.   c. decreased.
    b. doubled.   d. halved.

2. In the coils of a transformer, the motion of the flux   _D_
    is caused by the
    a. direct current.   c. moving secondary.
    b. rotating primary.   d. alternating current.

3. Energy is transferred from the primary to the secondary   _A_
    coils without a change in
    a. frequency.   c. current.
    b. voltage.   d. ampere-turns.

4. Transformer efficiency averages   _B_
    a. 79 percent.   c. 50 percent.
    b. 97 percent.   d. 100 percent.

5. A transformer has a primary coil rated at 150 volts and a   _C_
    secondary winding rated at 300 volts. The primary winding has
    500 turns. How many turns does the secondary winding have?
    a. 250   c. 1,000 $T_s = \dfrac{300 \times 500}{150} = 1000$
    b. 2,500   d. 10,000

6. A control transformer is a step-down transformer type.   _A_
    Compared to the secondary winding, the primary winding is
    a. larger in wire size.
    b. smaller in wire size.
    c. the same size as the secondary.
    d. connected to the load.

7. The current in the secondary winding   _A_
    a. is higher than the current in the primary.
    b. is lower than the current in the primary.
    c. controls the current in the secondary.
    d. controls the current in the primary.

B. Solve the following problems. $T_s = \dfrac{220 \times 100}{110}$

8. A 110/220-volt step-up transformer has 100 primary turns. How many turns does
    the secondary winding have? _____ $\dfrac{E_p}{E_s} = \dfrac{T_p}{T_s}$ _____ $\dfrac{100}{200}$ _____

    _____ 200 _____

9. A transformer has 100 primary turns and 50 secondary turns. The current in the
    secondary winding is 20 amperes. What is the current in the primary winding?
    $I_p = \dfrac{I_s \times T_s}{T_p}$ $I_p = \dfrac{20 \times 50}{100}$

    $I_p = \dfrac{1000}{100} = 10\,A$

10. What is the ratio of a transformer that has a secondary voltage of 120 volts when connected to a 2,400-volt supply? $\dfrac{E_p}{E_s}$ $\dfrac{2400}{120}$ $\dfrac{20}{1}$

11. A 7,200/240-volt step-down transformer has 1,950 primary turns. Determine the number of turns in the secondary winding.

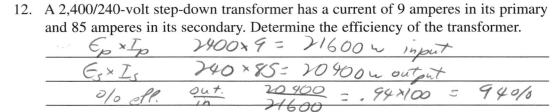

$$T_s = \frac{240 \times 1950}{7200} = \frac{468000}{7200} = 65$$

12. A 2,400/240-volt step-down transformer has a current of 9 amperes in its primary and 85 amperes in its secondary. Determine the efficiency of the transformer.

$\dfrac{E_p \times I_p}{E_s \times I_s}$ $2400 \times 9 = 21600 \, w \; input$

$240 \times 85 = 20400 \, w \; output$

$\%\; eff. \quad \dfrac{out.}{in} \quad \dfrac{20400}{21600} = .94 \times 100 = 94\%$

18

SINGLE-PHASE TRANSFORMERS

OBJECTIVES

After studying this unit, the student should be able to

- describe a single-phase, double-wound transformer, including its primary applications.
- diagram the series and parallel methods of coil connection for a double-wound transformer and for primary and secondary dual-voltage connections.
- define subtractive polarity and diagram the connections and markings for this polarity.
- define additive polarity and diagram the connections and markings for this polarity.
- list the steps in the AC polarity test for a single-phase transformer.
- demonstrate good electrical safety practices.
- describe an autotransformer, including its primary applications.
- identify primary taps.

A single-phase transformer usually has a core and at least two coils. The single-phase autotransformer has only one coil. The specifications for single-phase transformers vary greatly, and the applications of these transformers are unlimited.

THE DOUBLE-WOUND TRANSFORMER (ISOLATING AND INSULATING)

The double-wound transformer has a primary winding and a secondary winding. These windings are independently isolated and insulated from each other. A *shielded winding* transformer, on the other hand, is designed with a metallic shield between the primary and secondary windings, providing a safety factor by grounding. This prevents accidental contact between the windings under faulty conditions. The illustrations in Unit 17 show a double-wound transformer. The coils of double-wound transformers may be connected in several different arrangements.

Figure 18-1 shows popular arrangements of single-phase transformer windings. Shown in Figure 18-1(A), two single coils are used for specific step-down or step-up applications, including bell-ringing transformers, neon transformers, and component transformers for commercial equipment, such as automatic machines, switchgear, and

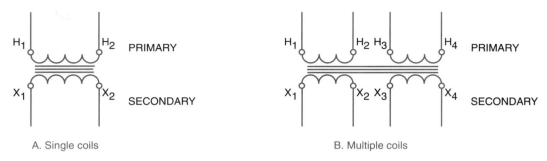

Figure 18-1   Coil arrangements for single-phase transformers.

other devices. Multiple coil primary and secondary windings, shown in Figure 18-1(B), are used in distribution transformers where dual voltage ratings are desired. Arrangements for voltage ratings of 2,400//120/240 or 240/480//120/240 are common.

Double-wound transformers separate or insulate the high transmission voltages from the typical consumer voltages of 115/230/460. The *National Electrical Code*® requires this type of transformer in all distribution circuits with the exception of those circuits assigned to autotransformers. Here, as in the *Code*®, the voltage considered shall be that at which the circuit operates, except for the examples given.

Polarity

A 460/115/230 transformer may be connected for two ratios: 460/115 or 460/230. To obtain the 460/115 ratio, the secondary coils are connected in parallel; the 460/230 ratio is achieved by connecting the secondary coils in series. To complete these connections, the polarity of the leads must be determined.

Figure 18-2 shows how the transformer secondary series and parallel coil connections are made. Note that instead of polarity indications such as (+, −), the coil leads are identified here by S (start) and F (finish), or by 1 (start) and 2 (finish) as in H_1, H_2, X_1, and X_2 in Figure 18-3.

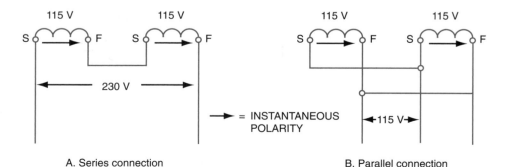

Figure 18-2   Series and parallel transformer and battery voltage connections.

The beginning or ending of a transformer coil is usually indicated by a tab placed on the lead by the technician in charge of the winding process. When the transformer is assembled, other markings often replace the original ones. Before final inspection, a polarity test must be made to be certain that the leads are marked correctly.

IDENTIFYING AN UNMARKED TRANSFORMER

Installed transformers often have missing or disfigured tabs. Every time a transformer is to be reconnected following repairs or for other reasons, the polarity of the leads must be checked.

Figures 18-3 and 18-4 illustrate two systems of marking polarity. In conventional usage, polarity refers to the induced voltage vector relationships of the transformer leads as they are brought outside of the tank. The American National Standards Institute (ANSI) standardized the location of these leads to obtain additive and subtractive polarity conditions. All high-voltage leads brought outside the case are marked H_1, H_2, and so forth,

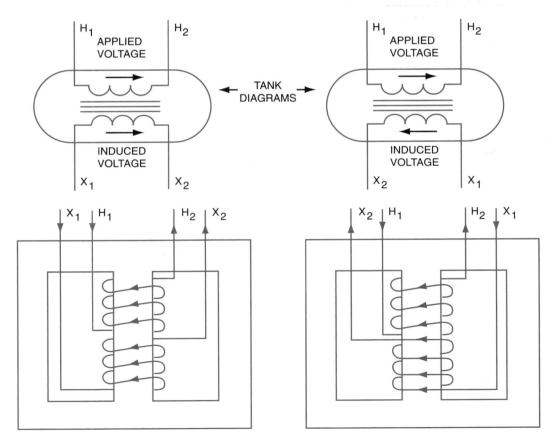

Figure 18-3 Subtractive polarity.   **Figure 18-4 Additive polarity.**

whereas the low-voltage leads are marked X_1 and X_2. The H_1 lead is located on the left side when facing the low-voltage leads. H_1 and X_1 are both positive at the same instant of time.

Subtractive Polarity. The tank diagram in Figure 18-3 shows the relationship of the induced voltages in the primary and secondary windings for the subtractive polarity condition. Transformers connected in this manner have the H_1 and X_1 leads located directly opposite each other. If H_1 and X_1 are connected together [as shown in Figure 18-5(A)], the voltage measured between H_2 and X_2 is less than the primary voltage. The induced voltage opposes the supply voltage and thus causes the secondary induced voltage to be subtracted from the primary voltage.

Additive Polarity. The tank diagram in Figure 18-4 shows the voltage relationship of the induced voltages for the additive polarity connection. When H_1 and X_2 are connected, the voltage across H_2 and X_1 is greater than the primary voltage. The measured voltages add up to the sum of the primary and secondary voltages as shown in Figure 18-5(B).

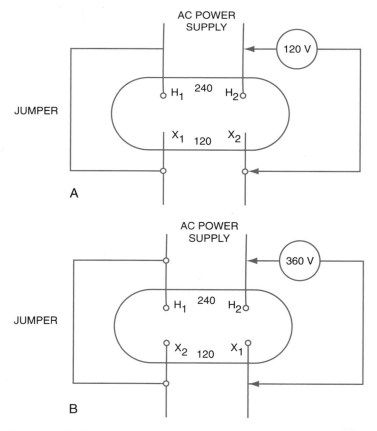

**Figure 18-5 (A) Test used to determine subtractive polarity transformer.
(B) Test used to determine additive polarity transformer.**

Transformers that are rated up to 200 kVA and have the value of the high-voltage winding equal to 8,660 volts or less are additive. All other transformers are subtractive.

Test for Polarity. Transformer coils often must be connected in series or parallel as in Figure 18-6. For these situations, the polarity of a transformer or any secondary coil can be found by making the connections shown in Figure 18-5(A) and (B). Connect the adjacent left, high- and low-voltage outlet leads facing the low-voltage side of the transformer. Apply a low-voltage supply to the primary and note the voltage between the adjacent right-hand, high- and low-voltage terminals.

- For *subtractive* polarity, the voltmeter reading (V) is less than the applied voltage. The voltage is the difference between the primary and secondary voltages, $E_p - E_s$ [Figure 18-5(A)].

- For *additive* polarity, the voltmeter reading (V) is greater than the applied voltage. The voltage is the sum of the primary and secondary voltages, $E_p + E_s$ [Figure 18-5(B)].

If the test shown in Figure 18-5(B) indicates additive polarity, the secondary leads inside the tank must be reversed at the bottom of the bushings to obtain a true subtractive polarity. If the transformer requires all additive polarity and the test indicates subtractive, reverse the secondary lead markers so that X_2 is located opposite H_1.

In all transformers, the H terminals are always the high-voltage terminals. The X terminals are always the low-voltage terminals. Either the H or X terminals can be

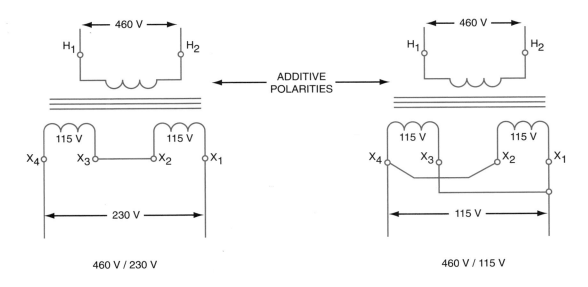

Figure 18-6 Single-phase transformer connections.

designated as the primary or the secondary, depending on which is the source and which is the load, and whether the transformer is used as a step-up or step-down transformer.

SINGLE-PHASE TRANSFORMER CONNECTIONS

Series Connection

If a 460/115/230-volt, single-phase transformer is to be connected to obtain 460/230 volts, the two secondary coils must be connected in series. The beginning and ending of each coil must be joined, as shown in Figure 18-6(A). The "start" of each coil is identified by an odd-numbered subscript.

Note: If the voltage is zero across X_1–X_4 after the series connections are complete, the coils are opposing each other (the polarity of one coil is reversed). To correct this situation, reverse one coil, and then reconnect and recheck the polarity.

Parallel Connection

To obtain 460/115 volts, the two secondary coils must be connected in parallel as shown in Figure 18-6(B). The polarity of each coil must be correct before making this connection. The parallel connection of two coils of opposite polarity will result in a short circuit and internal damage to the transformer.

Note: An indirect polarity check can be made by completing the series connection and noting the total voltage. As noted previously, zero voltage indicates opposite polarities. Reverse one coil to remedy the condition and then recheck overall polarity. Retag leads if necessary.

PARALLEL OPERATION OF SINGLE-PHASE TRANSFORMERS

Single-phase distribution transformers can be connected in parallel only if the voltage and percent impedance ratings of the transformers are identical. This information is found on the nameplates of large transformers. It is recommended that this rule be followed when making permanent parallel connections of all transformers.

TRANSFORMER PRIMARY TAPS

Taps are nothing more than alternative terminals that can be connected to more closely match the primary supply voltage. These taps are arranged in increments of 2.5 percent or 5 percent of the primary nominal voltage rating of the transformer (Figure 18-7). This provides a job-site adjustment to ensure that the transformer primary matches the supply voltage. The secondary will then produce the desired secondary voltage.

The voltage received from the power utility may be low or high. Because the transformer is a fixed voltage device, the output voltage is always in direct proportion to the input voltage. If the ratio is 2:1 and the supply voltage is 480 volts, the output will be 240 volts. If the primary voltage is 438 volts, the secondary is only 219 volts.

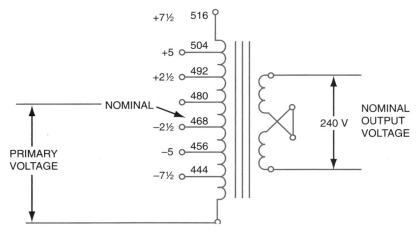

Figure 18-7 Primary terminal taps.

High and low voltages can have serious effects on different connected loads. Therefore, care must be taken to deliver a voltage as close as possible to the desired primary so that the secondary voltage will match the equipment nameplate voltages. Consistently high and low voltage problems can be solved by connecting the proper primary taps (Figure 18-8). If the voltage fluctuates consistently, tap changing is not the solution. A voltage regulating transformer is needed.

Figure 18-8 Tap connections used on three-phase power transformer.

REGULATION

A slight *voltage drop* at the secondary terminals from *no load* to *full load,* called regulation, is caused by resistance and reactance drops in the windings. Regulation is expressed as a percentage. Regulation of constant potential transformers is about 1 to 5 percent at the secondary terminals as follows:

$$\% \text{ E regulation} = \frac{\text{no load E} - \text{full load E}}{\text{full load E}} \times 100$$

Example: The secondary voltage of a transformer rises from 220 to 228 volts when the rated load is removed. What is the regulation of the transformer?

$$\% \text{ regulation} = \left(\frac{228 - 220}{220} \right) \times 100 = .036 \times 100 = 3.6\%$$

AUTOTRANSFORMER

Transformers with only one winding are called *autotransformers*. In an autotransformer, a part of the winding is common to both primary and secondary circuits. This is the most efficient type of transformer because a portion of one winding carries the difference between the primary and secondary currents. Figure 18-9 shows the current distribution in an autotransformer used in a typical lighting application. The disadvantage of an autotransformer is the fact that the use of only one winding makes it impossible to

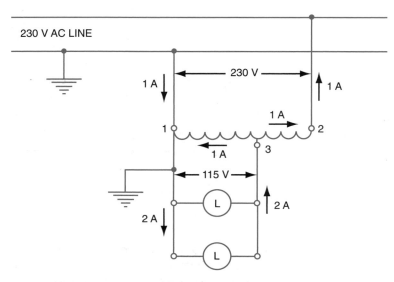

Figure 18-9 Autotransformer used for lighting.

Figure 18-10 Autotransformer used for motor-starting duty.

insulate the low-voltage section from the high-voltage distribution line. If the low-voltage winding opens when stepping down the voltage, the full-line voltage appears across the load. Autotransformers are used in some motor starters to reduce current draw from the line. See Figure 18-10 for an autotransformer motor starter.

According to the 2008 *National Electrical Code*® *Article 450,* autotransformers rated at 600 V or less shall have an individual overcurrent device connected in series with the ungrounded input conductor. If the primary current is less than 9 amps, the overcurrent rating shall not be more than 125 percent of the rated primary current. If the primary current is 9 amps or more, the overcurrent value can be the next standard larger size over 125 percent. An overcurrent device shall not be installed in series with the shunt winding (common winding to primary and secondary), with one exception noted.

DRY AND LIQUID-FILLED TRANSFORMERS

Dry transformers are used extensively for indoor installations. These transformers are cooled by air and are not encased in heavy tanks, such as those required for liquid-filled transformers. Dry transformers are used for bell-ringing circuits, current and potential transformers, welding transformers, and almost all transformers used on portable or small industrial equipment, or for separately derived systems within a facility.

Liquid-filled transformers consist of the core and coils immmersed in a tank of oil or other insulating liquid. Oil cooling is approximately fifteen times more effective than air cooling. Many distribution transformers designed for outdoor installation are liquid filled.

METHODS OF COOLING

The method selected to cool a transformer must not only maintain a sufficiently low average temperature, but must also prevent an excessive temperature rise in any portion of the transformers. In other words, the cooling medium must prevent the formation of "hot spots." For this reason, the working parts of the transformer are usually immersed in a high grade of insulating oil. Because the oil must be free of any moisture, the oil must be filtered to remove moisture as necessary. The electrical insulating value of the oil is checked periodically.

Duct lines are arranged within the transformer to provide for the free circulation of oil through the core and coils. The warmer and thus lighter oil rises to the top of the steel tank. The transformer core and windings are placed near the bottom of the tank. The cooler and heavier oil settles to the bottom of the tank. This natural circulation provides for better cooling (Figure 18-11).

Figure 18-11 An oil-filled power transformer with radiators.

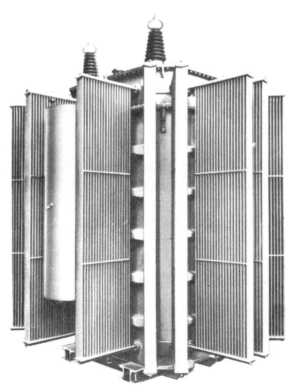

Forced Cooling

Several methods of removing heat from a transformer involve forced cooling. Cooling is acheived by using pumps to force the circulation of the oil or liquid, by forcing the circulation of air past the oil-filled radiators (Figure 18-12), or by immersing water-containing coils in the oil. Cold water circulating in the coils removes the heat stored in the oil. Figure 18-13 shows internal construction of an oil-filled transformer. Forced air movement by the use of fans is a common practice. Fans are generally controlled by thermostats (Figure 18-14).

APPLICATION

Single-phase transformers are suitable for use in a wide variety of applications, as shown by the examples in Figures 18-15, 18-16, 18-17, and 18-18.

Distribution transformers are usually oil filled and mounted on poles, in vaults, or in manholes.

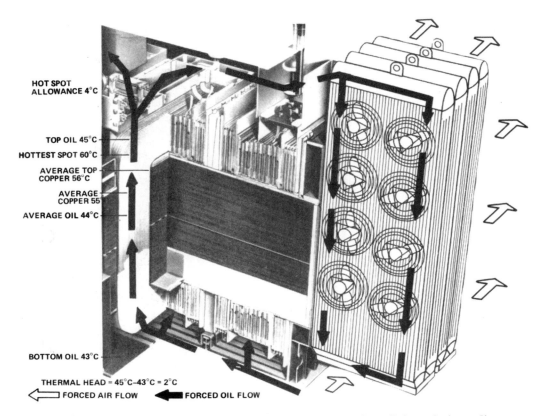

Figure 18-12 Cross section of a shell-form transformer showing oil-forced air cooling (FOA or FOA/FOA with typical temperature rises). (*Courtesy of Westinghouse Electric Corporation, Power Transformer Division*)

Figure 18-13 Assembly of a large three-phase, oil-filled, station-class power transformer, such as the one shown in Figure 18-11. (*Courtesy of McGraw-Edison Company, Power Systems Division*)

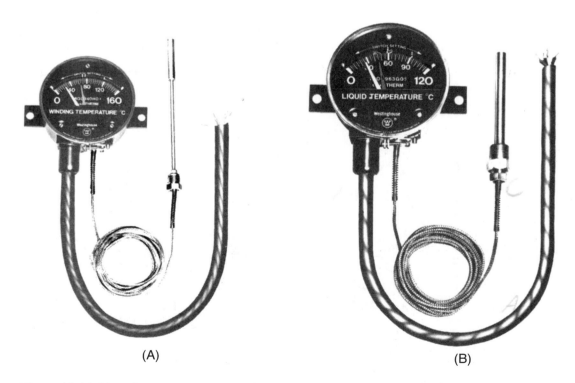

(A)   (B)

Figure 18-14 Transformer temperature indicators. (A) Winding temperature indicator. (B) Liquid temperature indicator. (*Courtesy of Westinghouse Electric Corporation, Power Transformer Division*)

Figure 18-15 Pole-mounted single-phase distribution transformer.

Figure 18-16 Cutaway view of a distribution transformer.

Figure 18-17 Control circuit transformer.

Figure 18-18 Transformer for electronic equipment.

Compensator starters are tapped autotransformers used for starting induction motors.

Instrument transformers, such as potential and current transformers, are made in indoor, outdoor, and portable styles used for metering.

Welding transformers provide a very low voltage to arc welding electrodes. Movable secondaries provide the varying voltage and current characteristics required.

Constant-current transformers are used for series street lighting where the current must be held constant with a varying voltage. The secondary is movable.

SAFETY PRECAUTIONS

Although there are no moving parts in a transformer, some maintenance procedures must be performed. For a general overhaul of an operating transformer or when an internal inspection is to be made, the transformer must be deenergized. Do not assume that the transformer is disconnected or rely on someone else to disconnect it; always check the transformer yourself. You must be sure that the fuses are pulled open or out and that the switch or circuit breaker is open on *both* the primary and secondary sides. After the transformer is disconnected, the windings should be grounded to discharge any capacitive energy stored in the equipment. This step also protects you while you are at work. Grounding is accomplished with a device commonly known as a "short and ground." This is a flexible cable with clamps on both ends. The ground end is clamped first, and then, using a hook stick, the other end is touched to the conductor. Do this with each leg on the primary and secondary sides. The phases are then shorted together and grounded for your protection.

The tank pressure should be relieved. This may be done by bleeding a valve or plug above the oil level. Any gas in the tank must be replaced with fresh air before a person enters the tank. The absence of oxygen in a tank will cause asphyxiation quickly and without warning. A second person should be on duty outside the transformer as a safety precaution whenever someone must enter the transformer. Be sure to follow Occupational Safety and Health Administration (OSHA) rules on hazardous entry and restricted work space. All tools should have safety cords attached with the other end tightly secured. All pockets in clothing should be emptied. Nothing must be allowed to fall into the tank. Great care must be exercised to prevent contacting or coming close to the electrical conductors and other live parts of the transformer unless it is known that the transformer has been deenergized. The tank and cooling radiators should not be touched until it is determined that they are adequately grounded (for both new and old installations).

SUMMARY

Single-phase transformers are used in a variety of applications. To connect them correctly, the electrician must know how the AC polarities are established at the transformer leads. Because the internal windings of the transformer are often not accessible, polarity checks must be made on the external leads. The transformer leads are marked according to the standards for additive or subtractive polarity. If the lead markings are not apparent or the voltages expected are not obtained, then the polarity must be tested. If the lead markings are known, the transformer coils may be connected in series or parallel to yield the desired voltage. Autotransformer connections or tap-changing transformers may be used to bring the voltage within the desired range. Much of a transformer's capability to operate satisfactorily depends on the capability to dissipate the heat produced in the windings. Many methods are used to dissipate this heat and keep the transformer from overheating and being destroyed.

ACHIEVEMENT REVIEW

Select the correct answers for each of the following statements and place the corresponding letter in the space provided.

1. Double-wound transformers contain a minimum of　　　　　　_C_
   a. one main winding.
   b. one main winding with two coils.
   c. a primary and a secondary winding.
   d. a primary and a double-wound secondary.

2. A transformer has subtractive polarity when the　　　　　　_C_
   a. two primary coil voltages oppose each other.
   b. two secondary coils have opposite polarities.
   c. X_1 lead is opposite the H_1 lead.
   d. X_2 lead is opposite the H_1 lead.

3. A transformer has additive polarity when the　　　　　　_D_
   a. two primary coils are in series.
   b. two secondary coils have aiding polarities.
   c. X_1 lead is opposite the H_1 lead.
   d. X_2 lead is opposite the H_1 lead.

4. Polarity should be tested before　　　　　　_C_
   a. energizing a transformer.
   b. checking the ratio.
   c. connecting the coils in series or parallel.
   d. connecting the load to the secondary.

5. A 440/110/220-volt, step-down transformer is connected　　　　　　_A_
   for 440/220 volts. Preliminary tests show that each secondary
   coil has 110 volts but the voltage across X_1–X_4 is zero.
   The probable trouble is that
   a. the voltages in the coils are equal and opposing.
   b. their ratings are equal.
   c. the load will divide in proportion to the capacities.
   d. the voltage drops at full load will be proportional to their respective loads.

6. The autotransformer may be used as a   _D_
   a. power transformer.
   b. potential transformer.
   c. current transformer.
   d. compensator motor starter.

7. Insulation of transformers may be classed in two groups: _B_
   a. double-wound and autotransformers.
   b. dry and oil-filled types.
   c. core and shell types.
   d. core and cross types.

8. Regarding cooling, transformers may be   _A_
   a. air- and oil-cooled.
   b. outdoor- and indoor-cooled.
   c. self- and forced-cooled.
   d. dry- and liquid-cooled.

9. Single-phase, double-wound transformers must be used for   _A_
   a. distribution and compensator starters.
   b. instrument and welding transformers.
   c. welding and dimming in theater lighting.
   d. constant current and reduced voltage motor starters.

10. For low voltage, the secondary of a single-phase transformer   _B_
    is connected:
    a. X_1 and X_4 to load, X_3 and X_2 together.
    b. X_1 to X_3 to load, X_2 to X_4 to load.
    c. H_1 to H_4 to load, H_3 to H_2.
    d. H_1 to H_2, X_1 to X_2.

11. A transformer in which part of the secondary is part of   _C_
    the primary is
    a. a series and parallel connection.
    b. a double-wound transformer.
    c. an autotransformer.
    d. an isolating transformer.

12. Parallel operation of single-phase transformers can be   _A_
    accomplished when the
    a. voltage and percentage impedance ratings are identical.
    b. voltage and current ratings are equal.
    c. cooling methods are identical.
    d. primary and secondary voltage ratings are equal.

13. Primary taps are designed to   _D_
    a. raise the voltage of the secondary.
    b. drain the oil.
    c. lower the voltage of the secondary.
    d. raise or lower the voltage of the secondary.

14. A slight voltage drop at the secondary terminals from no   _B_
    load to full load is called
    a. reactance.   c. percentage.
    b. regulation.   d. taps.

15. When working in a large transformer, the electrician should   _B_
    a. ventilate it first.
    b. short and ground all windings.
    c. secure all tools and empty pockets.
    d. all of the above

16. When preparing to work on an oil-filled transformer,   _A_
    a. bleed the tank pressure.
    b. disconnect the supply voltage and load.
    c. disconnect all other connections.
    d. check the disconnect switches yourself.

19

THE SINGLE-PHASE, THREE-WIRE SECONDARY SYSTEM

OBJECTIVES

After studying this unit, the student should be able to

- diagram the connections for a single-phase, three-wire secondary system.
- list the advantages of a three-wire service.
- describe what occurs when the neutral of a three-wire secondary system opens.
- explain why there is less copper loss for a three-wire system.

Most homes are wired for three-wire service. Because electric ranges and clothes dryers are designed for three-wire operation, any home to be provided with these appliances must have three-wire service. The three wires terminate in the residence at the load center panel so that most individual circuits carried through the house are at 120 volts, thus eliminating the dangers with 240-volt circuits.

The double-wound transformer is used as the source for three-wire secondary distribution. One of the important advantages of a transformer is its capability to provide a three-wire circuit from the low-voltage secondary. A step-down transformer with a 2,400/240/120-volt rating is commonly used in residential installations.

The advantages of using three-wire service in general distribution systems include (1) a reduction in the cost of main feeders and subfeeders, (2) the provision of 120-volt service for normal lighting circuits and 240-volt service for power and motor loads, and (3) the conservation of electrical energy by reducing wattage loss in transmission.

Figure 19-1 is a schematic of a typical three-wire system. The secondary coils are connected in series and each coil is rated at 120 volts. The junction N between the two secondary coils is usually grounded. This precaution provides some protection to a person who may come into contact accidentally with a transformer that has faulty insulation. The line wire carried from this junction to the several loads is known as the *neutral* or *grounded conductor*. The neutral wire generally carries less current than wires L_1 and L_2, except when the load is on one side only, that is, L_1 to N or L_2 to N. The 240-volt motor load does not affect the current flowing in the neutral wire. The neutral is carried through the system as a solid conductor (not fused or switched). If the neutral opens, and the loads in the 120-volt circuits are greatly unbalanced, then these 120-volt circuits will

169

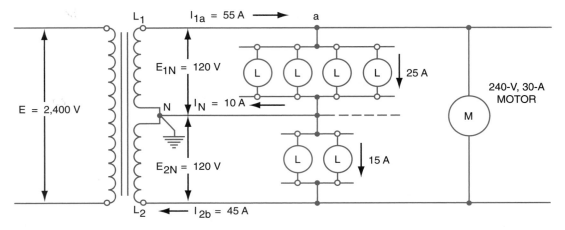

Figure 19-1 Schematic diagram of a three-wire system supplied from a single-phase transformer.

be subjected to approximately 240 volts. The neutral is designed to carry not only the unbalanced current in the two 120-volt circuits, but also the entire load on any one side should all the load on the other side be cut off completely. This latter situation can occur if a fuse or circuit breaker suddenly opens either line. Figure 19-1 shows the current distribution for the loads indicated.

OPEN NEUTRAL

As an example to show what occurs when the neutral of a three-wire system opens, assume that the lighting load in Figure 19-1 is a pure resistive load.

Thus, the group of four lamps has a resistance of $\dfrac{E}{I} = \dfrac{120}{25} = 4.8$ ohms, and the group of two lamps has a resistance of $\dfrac{E}{I} = \dfrac{120}{15} = 8$ ohms.

With the neutral open, these two groups combine as a series circuit with a resistance of 12.8 ohms connected across 240 volts. The current flow through this series circuit follows:

$$\frac{E}{R} = \frac{240}{12.8} = 18.75 \text{ amperes}$$

Then, according to the laws for a series circuit, the voltage across the 8-ohm group (two lamps) is equal to

$$I \times R = 18.75 \times 8 = 150 \text{ volts}$$

and the voltage across the 4.6-ohm group (four lamps) is equal to

$$I \times R = 18.75 \times 4.8 = 90 \text{ volts}$$

(Remember that in a series circuit, the highest voltage appears across the highest value of resistance.) The lamps would probably burn out with this open neutral.

Sample Problem

Referring to Figure 19-1, assume that the upper 120-volt load is 25 amperes, the lower load is 15 amperes, and the motor load is 30 amperes. If the power factor in all cases is unity (1), calculate the current

1. in line 1a,

2. in line 2b, and

3. in the neutral line N.

In addition, determine the power delivered

4. by transformer coil 1N,

5. by transformer coil 2N, and

6. by the primary coil.

Finally, calculate the current

7. in the primary coil.

Solution

1. I_{1a} = 25 + 30 = 55 amperes

2. I_{2b} = 15 + 30 = 45 amperes

3. I_N = 25 − 15 = 10 amperes

4. P_{1N} = 55 × 120 = 6,600 watts

5. P_{2N} = 45 × 120 = 5,400 watts

6. P_{pri} = 6,600 + 5,400 = 12,000 watts

7. I_{pri} = 12,000/2,400 = 5 amperes

The distribution transformers used in industrial plants or network substations for three-wire secondary systems are usually mounted on poles (Figure 19-2) or in transformer vaults. This type of transformer is equipped with three low-voltage bushings and the series connection is made inside the tank. The lower lines constitute the secondary three-wire systems.

Figure 19-2 Pole-top transformers used for distribution voltages; three low-voltage bushings are shown here.

ECONOMICS OF THE THREE-WIRE SYSTEM FOR FEEDERS AND BRANCH CIRCUITS

Using the three-wire system of the previous problem as an example, the total load transmitted over the three wires is 12,000 W or 12 kW at a power factor of 100 percent. It is assumed that the motor load is provided with power factor correction. If single conductor-type TW wire is used from the transformer to the load, the following sizes are required:

Line 1 (55 amperes): No. 6 TW

Neutral (0.70 × 55 = 38.5): No. 8 TW

Line 2 (55 amperes): No. 6 TW

Although No. 8 TW wire is the actual size permitted for the neutral, a substitution of a No. 6 TW wire can be made so that three 6 lines are provided to simplify the installation.

If a two-wire distribution system is used for the same load, the total current is 11,500/120 = 96 amperes. Two 1 lines are required. If the transmission distance is 100 feet, then a comparison can be made of the weights of copper wire required for the two systems.

Three-Wire System

For a No. 6 TW line, the weight per 100 feet = 11.5 lb. Therefore, for three No. 6 TW lines, the total weight = $3 \times 11.5 = 34.5$ lb.

Two-Wire System

For a 1 line, the weight per 100 feet = 33 lb. For two 1 lines, the total weight = $2 \times 33 = 66$ lb. Therefore, for the same load, the three-wire system uses less copper ($66 - 34.5 = 31.5$ pounds less) than the two-wire system.

A similar conclusion can be reached by consulting a manufacturer's price list and noting the lower prices for smaller conductors. The copper losses in the line are also considerably less for a three-wire system for several reasons: the motor power is transmitted at a higher voltage requiring less current for a given load; the neutral carries no current when the two lighting circuits are balanced; and the copper losses are much less because less wire and current are required. These line losses are of two types: (1) voltage drop (IR) and (2) wattage loss (I^2R).

SUMMARY

The three-wire single-phase system is the most common residential electrical service. The three wires constitute a single phase of AC delivered to the home. The transformer secondary coil is tapped at the center point and grounded at that point to establish a ground reference. The power can then be divided into two 120-volt supplies and can also be used as a 240-volt single phase for higher power consumption appliances. Care must be taken to solidly ground the neutral center-tapped point because an open neutral conductor can cause severe damage.

ACHIEVEMENT REVIEW

1. Cite two reasons why power companies must supply three-wire service to residential occupancies.

   _____

   _____

   _____

2. How are the two secondary coils of a distribution transformer connected for three-wire service? _____

   _____

   _____

3. What are three advantages of a three-wire service as compared to a conventional two-wire service? _____

    _____

    _____

    _____

    _____

4. Why must the neutral line be left unfused? _____

    _____

    _____

    _____

5. How many circuits are provided in a three-wire secondary system? _____

    _____

6. What is the voltage rating of each circuit in question 5?_____

    _____

    _____

    _____

    _____

    _____

Questions 7 through 9 are based on Figure 19-3: a three-wire system has one lighting load of 40 amperes, one lighting load of 20 amperes, and a 240-volt motor load of 30 amperes.

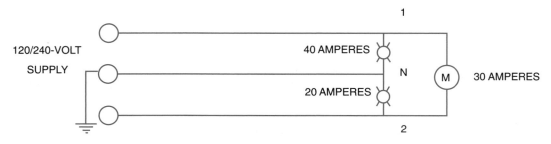

Figure 19-3 Three-wire system for questions 7 through 9.

7. What is the current load in lines 1 and 2?

    _____   _____

    (1)   (2)

8. What is the current in N?   _____

9. If the neutral is open, indicate the voltages of the lighting circuits. Show your
    work.

    1N _____   2N _____

10. A three-wire, 120/240-volt circuit supplies the following:
       One 120-volt, 10-watt lamp to line 1 and neutral, and
       one 120-volt, 120-watt TV set to line 2 and neutral. (See Figure 19-4.)

    If the neutral opens while both the lamp and the TV set are operating, what will
    be the voltages at the lamp and the TV set? (Assume a power factor of unity.)

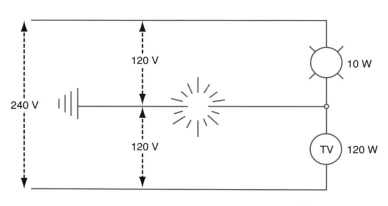

Figure 19-4 Three-wire system for question 10.

                                    (lamp voltage)   _____
                                    (TV voltage)   _____

11. A manufacturer uses motors larger than 1 hp on 120 volts. Why does this require
    a three-wire secondary system? _____

    _____

    _____

    _____

12. Theoretically, how much horsepower is on the unbalanced lines in Figure 19-5?
    L₁ _____ L₂ _____ Neutral _____

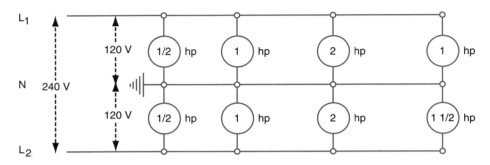

Figure 19-5 Isolation transformer connection for question 12.

20

SINGLE-PHASE TRANSFORMERS CONNECTED IN DELTA

OBJECTIVES

After studying this unit, the student should be able to

- explain, with the aid of diagrams, how single-phase transformers are connected in a three-phase, closed delta-delta arrangement.
- describe the relationships between the voltages across each coil and across the three-phase lines for both the input (primary) and output (secondary) of a delta-delta transformer bank.
- list the steps in the procedure for checking the proper connection of the secondary coils in the closed-delta arrangement, including typical voltage readings.
- describe how a delta-delta connected transformer bank can provide both a 240-volt, three-phase load and a 120/240-volt, single-phase, three-wire load.
- describe, using diagrams, the open-delta connection and its use.
- identify primary taps for the three-phase connection.

Most electrical energy is generated by three-phase, alternating-current generators. Three-phase systems are used to transmit and distribute the generated electrical energy. The voltage on three-phase systems often must be transformed, either from a higher value to a lower value or from a lower value to a higher value.

Voltage transformation on three-phase systems can be obtained with the use of three single-phase transformers (Figure 20-1). These transformers can be connected in several ways to obtain the desired voltage values.

A possible connection pattern that the electrician is often required to use for the three single-phase transformers is the closed-delta connection.

Another connection pattern that is commonly used is the open-delta or V connection that requires only two transformers to transform voltage on a three-phase system.

CLOSED-DELTA CONNECTION

When three single-phase coils are connected so that each coil end is connected to the beginning of another coil, a simple closed-delta system is formed (Figure 20-2).

Figure 20-1 Three large single-phase, station-class, oil-filled power transformers.
(*Courtesy of McGraw-Edison Company, Power Systems Division*)

Figure 20-2 Simple delta connection.

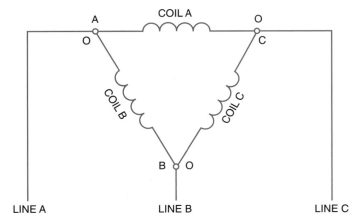

When the three coils are marked Coil A, Coil B, and Coil C, the end of each of the three coils is marked with the letter O. The beginnings of the coils are marked A, B, and C. Note that each coil end is connected to another coil beginning. Each of the three junction points ties to a line lead feeding a three-phase system.

If three single-phase transformers are to be used to step down 2,400 volts, three phase to 240 volts, three phase, a closed-delta connection is used. Each of the three

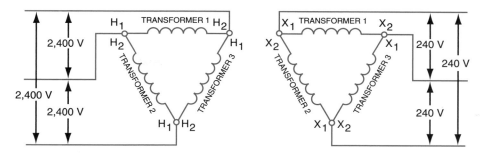

Figure 20-3 Elementary diagram of delta-delta transformer connections.

transformers is rated at 2,400 volts on the high-voltage side and 240 volts on the low-voltage side (Figure 20-3).

CONNECTING THE DELTA

The transformer leads on the high-voltage side of each single-phase transformer are marked H_1 and H_2. The leads on the low-voltage side of each single-phase transformer are marked X_1 and X_2.

To connect the high-voltage primary windings in the closed-delta pattern to a three-phase source, the three windings are connected as follows: in making the connection, the end of one primary winding is connected to the beginning of the next primary winding. In Figure 20-3, H_1 is the beginning of each coil and H_2 is the end of each coil. Thus, each primary winding end H_2 is connected to the beginning H_1 of another primary winding. A three-phase line wire is also connected at each junction point H_1–H_2. Note that the primary winding of each transformer is connected directly across the line voltage. This means that delta-connected transformers must be wound for the full-line voltage. In Figure 20-3, each of the three line voltages is 2,400 volts, and the primary winding of each transformer is also rated at 2,400 volts. After the high-voltage primary connections are made, the three-phase, 2,400-volt input may be energized. It is not necessary to make polarity tests on the input side.

The next step is to connect the low-voltage output or secondary windings in the closed-delta pattern. The secondary winding leads are marked X_1 for the beginning of each coil and X_2 for the end of each coil. In making the connections on the secondary, the following procedure must be followed:

1. Check to see that the voltage output of each of the three transformers is 240 volts.

2. Connect the end of one secondary winding with the beginning of another secondary winding (Figure 20-4).

The voltage across the open ends shown in Figure 20-4 should be the same as the output of each transformer or 240 volts. If one of the transformers has its secondary winding connections reversed, the voltage across the open ends will be $1.73 \times 240 = 415$ volts.

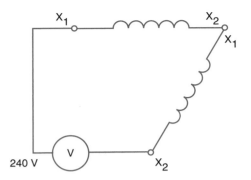

Figure 20-4 A voltmeter is used to check correct connections.

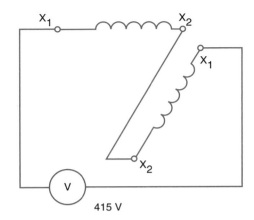

Figure 20-5 A voltmeter reading indicates incorrect connections.

Figure 20-5 illustrates an incorrect connection that must be changed so that it is the same as the connection shown in Figure 20-4.

Figure 20-6 illustrates the correct connections for the secondary coil of the third transformer. The voltage across the last two open ends should be zero if all the transformers are connected as shown. If the voltage is zero across the last two open ends, they may be connected together. A line lead is then connected at each of the three junction points X_1–X_2. These three wires are the 240-volt, three-phase output. Note that each of the three line voltages and each of the three transformer output voltages is equal to 240 volts.

When the secondary winding of the third transformer is reversed, the voltage across the last two open leads would be 240 + 240 = 480 volts.

Figure 20-7 illustrates the incorrect connection that results in a reading of 480 volts. The connections on the third transformer secondary must be reversed.

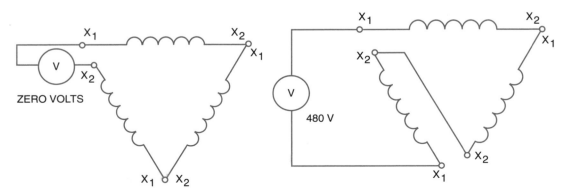

Figure 20-6 Voltmeter reading indicates correct connections.

Figure 20-7 A voltmeter reading indicates incorrect connections.

Caution: Never complete the last connection if there is a voltage difference greater than zero. If the connections are correct, this potential difference is zero. Observe safety precautions. Deenergize the primary while making connections.

When three transformers are connected with their primary windings in delta and their secondary windings in delta, the total connection is called a *delta-delta* (Δ-Δ) *connection*. The first delta symbol indicates the connection method of the primary windings, and the second delta symbol shows how the secondary windings are connected. When two or three single-phase transformers are used to step down or step up voltage on a three-phase system, the group is called a *transformer bank*.

Figure 20-8 is another way of showing the closed-delta connection first illustrated in Figure 20-3. By tracing through the connection, it can be seen that the high-voltage and low-voltage windings are all connected in the closed-delta pattern. This type of transformer diagram is often used by the electrician.

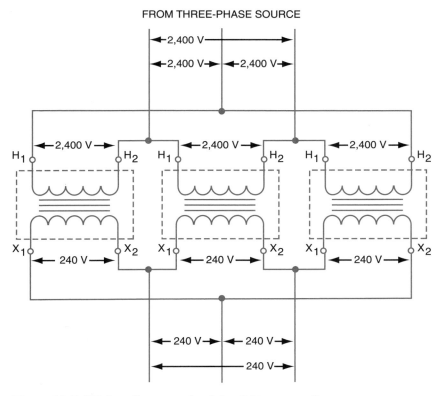

Figure 20-8 Wiring diagram of a delta-delta connection.

VOLTAGE AND CURRENT

In any closed-delta transformer connection, two important facts must be kept in mind.

- The line voltage and the voltage across the transformer windings are the same. A study of any delta connection shows that each transformer coil is connected directly across two line leads; therefore, the line voltage and the transformer coil voltage must be the same.
- The line current is greater than the coil current in a delta-connected transformer bank. The line current is equal to 1.73 × coil current. A study of a closed-delta transformer connection shows that each line lead is fed by two transformer coil currents that are out of phase and thus cannot be added directly.

In the arrangement shown in Figure 20-9, the coil current in each transformer secondary is 10 amperes. The line current, however, is 1.73 × 10 or 17.3 amperes. Because the coil currents are out of phase, the total current is *not* 10 + 10, or 20, amperes. Rather, the total current is a resultant current in a balanced closed-delta system and is equal to 1.73 × coil current (1.73 equals the square root of three).

Three single-phase transformers of the same kilovolt-ampere (kVA) capacity are used in almost all delta-delta connected transformer banks used to supply balanced three-phase industrial loads. For example, if the industrial load consists of three-phase motors, the current in each line wire is balanced. To determine the total kVA capacity of the entire delta-delta connected transformer bank, add the three transformer kVA ratings. Thus, if each transformer is rated at 50 kVA, the total kVA is 50 + 50 + 50 = 150 kVA.

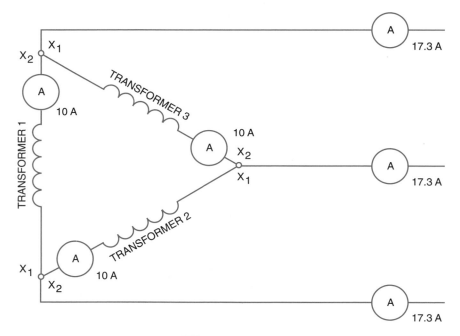

Figure 20-9 Line current is $\sqrt{3}$ times the coil current in a delta connection.

POWER AND LIGHTING SERVICE FROM A DELTA-DELTA CONNECTED TRANSFORMER BANK

A delta-delta connected transformer bank, with one transformer secondary center tapped, may be used to feed two types of load: (1) a 240-volt, three-phase industrial power load, and (2) a 120/240-volt, single-phase, three-wire lighting load.

The single-phase transformer that is to supply the single-phase, three-wire lighting load is usually larger in size than the other two transformers in the bank. This takes care of the additional lighting load placed here. A tap must be brought out from the midpoint of the 240-volt, low-voltage winding so that the 120/240-volt, single-phase, three-wire service can be obtained. Many transformers are designed with the low-voltage side consisting of two 120-volt windings. These windings can be connected in series for 240 volts, in parallel for 120 volts, or in series and center tapped, with the tap brought out to give 120/240-volt service.

Figure 20-10 illustrates three single-phase transformers connected as a delta-delta transformer bank. Each transformer has two 120-volt, low-voltage windings. These 120-volt windings are connected in series to give a total output voltage of 240 volts for each transformer. The connection scheme for the high-voltage input or primary windings is closed delta. The low-voltage output or secondary windings are also connected in the closed-delta pattern to give three-phase, 240-volt service for the industrial power load. Note in Figure 20-10 that the middle transformer is feeding the single-phase, three-wire, 120/240-volt lighting load. This center transformer has a middle tap on the secondary (output) side to give 120/240-volt service. Also note that this tap feeds to the grounded neutral wire. Figure 20-11 shows a one-line diagram representation.

The three-phase, 240-volt industrial power system is also connected to the transformer bank shown in Figure 20-11. A check of the connections shows that both lines A and C of the three-phase, 240-volt system have 120 volts to ground. Line B, however, has 208 volts to ground ($120 \times 1.73 = 208$). This situation is called the *high phase*.

Article 110.15 of the *National Electrical Code*® requires that the phase be marked orange in color and be placed in the middle in a switchboard or panelboard as "B" phase.

Caution: The high-phase situation can be a serious hazard to human life as well as to any 120-volt equipment connected improperly between the high phase and neutral. When the voltage to ground exceeds 250 volts on any conductor in any metal raceway or metal-sheathed cable, the *National Electrical Code*® requires special bonding protection.

For example, if rigid conduit is used to connect the services, there must be two locknuts. One locknut is used outside and one inside any outlet box or cabinet. The regular conduit end bushing must also be used to protect the insulation on the wires in the conduit. Where the conductors are above a given size, this conduit bushing must be the insulating type or equivalent, according to the *National Electrical Code*® in the section on cabinets.

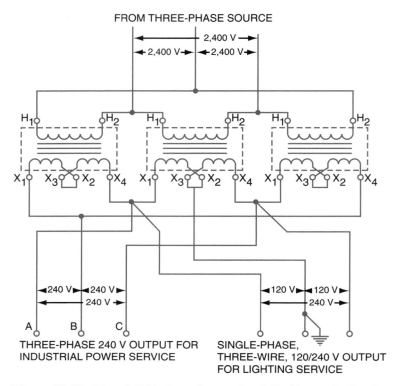

FROM THREE-PHASE SOURCE

THREE-PHASE 240 V OUTPUT FOR
INDUSTRIAL POWER SERVICE

SINGLE-PHASE,
THREE-WIRE, 120/240 V OUTPUT
FOR LIGHTING SERVICE

**Figure 20-10 Closed-delta transformer bank feeding a single-phase,
three-wire lighting load and a three-phase, three-wire power load.**

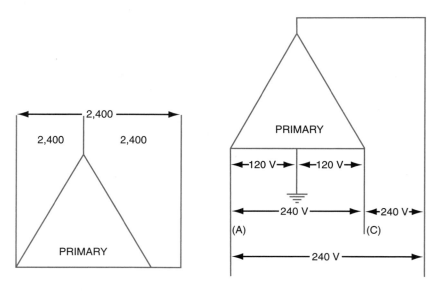

**Figure 20-11 One line diagram of Δ-Δ transformer with three-phase,
four-wire secondary.**

Note that for ungrounded circuits, the greatest voltage between the given conductor and any other conductor of the circuit is considered the voltage to ground.

OPEN-DELTA OR V CONNECTION

A three-phase transformation of energy is possible using only two transformers. This connection arrangement is called the open-delta or V connection. The open-delta connection is often used in an emergency when one of the three transformers in a delta-delta bank becomes defective. When it is imperative that a consumer's three-phase power supply be restored as soon as possible, the defective transformer can be cut out of service using the open-delta arrangement.

The following example shows how the open-delta connection can be used in an emergency. Three 50-kVA transformers, each rated at 2,400 volts on the high-voltage winding and 240 volts on the low-voltage winding, are connected in a delta-delta bank (Figure 20-12). This closed-delta bank is used to step down a 2,400-volt, three-phase input to a 240-volt, three-phase output to supply an industrial consumer. Suddenly, the three-phase power service is interrupted because lightning strikes and damages one of the transformers. The service must be restored immediately. This situation is shown in Figure 20-13.

Transformer 3 is the defective transformer. If all the leads of the damaged transformer are disconnected, the closed-delta bank automatically becomes an open-delta transformer bank.

The schematic diagram of this open-delta connection is shown in Figure 20-14. Note that with the one transformer removed, the triangular coil arrangement is open on one side. Because the schematic diagram resembles the letter V, this arrangement is also called the V connection.

Although it appears that the total kVA of the open-delta bank should be

Figure 20-12 Three single-phase transformers used to create three-phase distribution system.

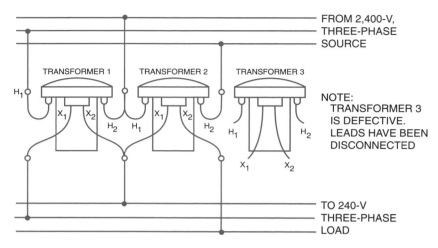

Figure 20-13 Open-delta connection.

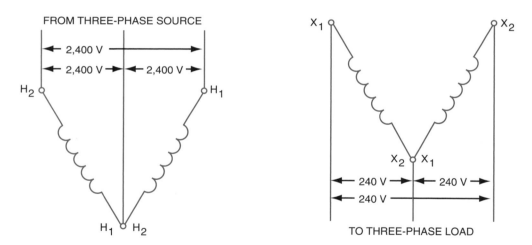

Figure 20-14 Schematic diagram of the open-delta or V connection.

two-thirds that of a closed-delta bank, the actual kVA rating of an open-delta bank is only 58 percent of the capacity of a closed-delta bank. The reason for this is that the currents of the two transformers in the open-delta connection are out of phase, resulting in the total available capacity of the open-delta bank being only 58 percent instead of 66.7 percent.

In the open-delta example, three 50-kVA transformers are connected in a delta-delta bank. This gives a total kVA capacity of 50 + 50 + 50 = 150 kVA for the closed-delta bank. When one transformer is disconnected, the transformer bank changes to an open-delta configuration, and the total kVA capacity now is only 58 percent of the original closed-delta capacity.

$$150 \times 0.58 = 87 \text{ kVA}$$

In some situations, an open-delta bank of transformers is installed initially. The third transformer is added when the increase in industrial power load on the transformer bank warrants the addition. When the third transformer is added to the bank, a closed-delta bank is formed.

When two transformers are installed in an open-delta configuration, the total bank capacity can be found by the use of the following procedure.

1. Add the two individual transformer kVA ratings. (For the problem given, the single-phase transformers are rated at 50 kVA.)

$$50 + 50 = 100 \text{ kVA}$$

2. Multiply the total kVA value by 86.5 percent. This gives the total kVA capacity of the open-delta transformer bank.

$$100 \times 86.5\% = 87 \text{ kVA}$$

Therefore, an open-delta bank has a kVA capacity of 58 percent of the capacity of a closed-delta bank; an open-delta bank has a kVA capacity of 86.5 percent of the capacity of two transformers.

Another way to explain the reduced percentages of output kVA is to use the rated voltages and currents. In an open-delta pattern, there is no vector addition of the current at the junction point; the line current is equal to the coil current. Just as in the closed-delta pattern, the open-delta voltage at the lines is the same as the coil voltages. The results can be seen in the following example: If each of the transformers are rated at 50 kVA and the secondary voltages are 240 volts, then the coil current of each transformer is 50,000/240 = 208 A. In an open-delta pattern, Line I equals Coil I and Line E equals Coil E. The three-phase capacity of two transformers connected open-delta is

$$\text{Line E} \times \text{Line I} \times 1.73 = 86.5 \text{ kVA}$$

This is the same as 86.55 percent of the two kVA added. This is also the same as 58 percent of the original 150 kVA or 1.73 times the single 50 kVA.

THREE-PHASE TRANSFORMERS WITH PRIMARY TAPS

Some plant distribution transformers are preassembled and wired at the factory into a three-phase bank in a single enclosure or as a single unit. These assemblies consist of three single-phase transformers in one enclosure, usually the dry, air-cooled type. Some have primary tap terminals so that the supply voltage can be matched more closely (Figure 20-15). The electrician must make the adjustment on the job until the primary of

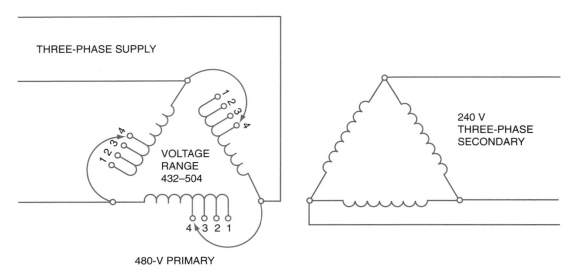

THREE-PHASE SUPPLY

VOLTAGE
RANGE
432–504

480-V PRIMARY

240 V
THREE-PHASE
SECONDARY

Figure 20-15 Tap connections for a three-phase bank.

the transformer matches the measured supply voltage. The secondary will then produce the desired voltage to achieve a closer match of the equipment nameplate voltages. Utilities do not always supply the desired accurate voltages. There may also be a voltage drop within the plant.

When using taps on a three-phase transformer or bank of transformers, it is important that the same taps on each of the three primaries be connected in the same position on each coil. (See the "Transformer Primary Taps" section in Unit 18.) The following problems may result if the taps are not connected properly:

- The output voltage on each of the three secondary voltages will not be the same. This will produce high unbalanced currents that will cause overheating of induction motors.

- An undesirable circulating current will create a "false load" condition if the transformer is connected delta-delta.

Taps are used for consistently high or low voltages. They are not used with voltages that fluctuate or vary frequently.

SUMMARY

Single-phase transformers are often used to create various patterns to supply three-phase loads. One pattern is the closed-delta pattern. In this pattern, the line voltage is the same as the phase voltage, but the current on the lines is 1.73 times the coil current. Be sure the coil leads are marked correctly and double-check the connection procedures before energizing the delta transformer bank.

Single-phase transformers can be connected in an open-delta pattern to provide a reduced capacity power supply to a system if one of the phase transformers fails. Single-phase transformers connected in the closed- or the open-delta pattern do not need to be the same kVA rating. Often, one transformer is larger if the system is to supply three-phase delta and some single-phase, three-wire systems. If the proper nominal voltage is not available at the primary of the transformer, primary taps may be needed to bring the voltage back to the proper level.

ACHIEVEMENT REVIEW

1. What is one practical application of single-phase transformers connected in a delta-delta configuration? _____

   _____

   _____

   _____

   _____

2. What simple rule must be followed in making a delta connection? _____

   _____

   _____

   _____

3. Make a connection diagram for three single-phase transformers connected in a closed-delta scheme. This transformer bank is used to step down 2,400 volts, three-phase to 240 volts, three-phase. Each transformer is rated at 50 kVA, with 2,400 volts on the high-voltage winding and 240 volts on the low-voltage winding. Mark leads H_1, X_1, and so forth. Show all voltages.

4. What is the total kVA capacity of the closed-delta transformer bank in question 3?

   _____

5. What is one practical application of an open-delta transformer bank?

   _____

   _____

   _____

   _____

   _____

   _____

6. Make a connection diagram of two single-phase transformers connected in open delta. Each transformer is rated at 10 kVA, with 4,800 volts on the high-voltage winding, and 240 volts on the low-voltage winding. This bank of transformers is to step down 4,800 volts, three phase to 240 volts, three phase. Mark leads H_1, X_1, and so forth. Show all voltages. Calculate the total kVA capacity of this open-delta transformer bank.

7. What problems are likely to result if taps are not connected properly on a three-phase transformer bank?

   _____

   _____

   _____

   _____

   _____

   _____

   _____

21

SINGLE-PHASE TRANSFORMERS CONNECTED IN A WYE INSTALLATION

OBJECTIVES

After studying this unit, the student should be able to

- diagram the simple wye connection of three transformers.
- list the steps in the procedure for the proper connection and checking of the primary and secondary windings of three single-phase transformers connected in a wye arrangement.
- state the voltage and current relationships for wye-connected, single-phase transformers.
- describe how the grounded neutral of a three-phase, four-wire, wye-connected transformer bank maintains a balanced voltage across the windings.
- state how the kVA capacity of a wye-wye connected transformer bank is obtained.

Voltage transformation on three-phase systems can also be accomplished using wye-connected, single-phase transformers (Figure 21-1). To avoid errors when wye connecting single-phase transformers, a systematic method of making the connections should be used. The electrician should know the basic voltage and current relationships common to this type of connection.

Figure 21-1 Single-phase, round-coil transformer. (*Courtesy of McGraw-Edison Company, Power Systems Division*)

FUNDAMENTAL WYE CONNECTION

A simple wye system is formed by arranging three single-phase coils so that one end of each coil is connected at a common point (Figure 21-2). Note that when these connections are shown in a schematic diagram, they resemble the letter Y (written as "wye"). This configuration is also known as a "star" connection.

As an example, Figure 21-3 shows the wye-wye connection of three single-phase transformers to step down a three-phase input of 4,152 volts to a three-phase output of 208 volts. Each transformer must be voltage rated for its applications. The H_2 leads or primary winding ends of the transformer are connected together. The beginning or H_1 lead of each transformer is connected to one of the three line leads.

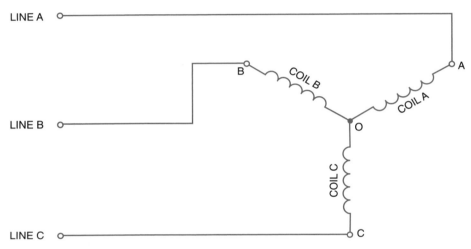

Figure 21-2 Simple wye connection.

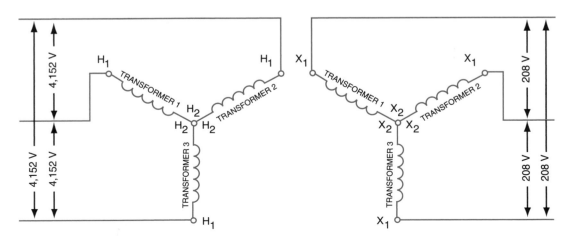

Figure 21-3 Elementary diagram of wye-wye connected transformer bank.

Two of the primary windings are connected across each pair of line wires. Each transformer primary winding is rated at 2,400 volts, and the actual voltage applied to each of these three windings is 2,400 volts. Note that the potential across each pair of line leads is 4,152 volts and not the 2,400 + 2,400 = 4,800 volts that might be expected because of the connection of two coils.

The value of 4,152 volts arises from the fact that the voltage applied to each of the primary windings is out of phase with the voltages applied to the other primary windings. As a result, these winding voltages cannot be added directly to obtain the line voltage. Rather, the line voltage is equal to 1.73 × coil voltage. Therefore, for the input side of the transformer bank in Figure 21-3 where the voltage on the primary winding of each transformer is 2,400 volts,

$$\text{Line voltage} = 1.73 \times \text{coil voltage}$$
$$= 1.73 \times 2,400$$
$$= 4,152 \text{ volts}$$

If the coil voltage must be checked and the line voltage is known, the same value of 1.73 can be used. For this situation, the coil voltage is obtained by dividing the line voltage by 1.73.

$$\text{Coil voltage} = \frac{\text{line voltage}}{1.73}$$
$$= \frac{4,152}{1.73}$$
$$= 2,400 \text{ volts}$$

Thus, wye-connected transformer banks have only 58 percent $\left(\frac{1}{1.73} = .58\right)$ of the line voltage applied to each of the three transformer windings. After the high-voltage primary connections are completed, the three-phase, 4,152-volt input may be energized. It is not necessary to make any polarity tests on the input (primary) side.

POLARITY TEST FOR UNMARKED AND NEW TRANSFORMERS

The next step is to connect the low-voltage output (secondary) windings in wye (Figure 21-3). Follow this procedure when making the secondary connections:

1. Check to see that the voltage output of each of the three transformers is 120 volts (for this example).

   Caution: Deenergize all circuits before making connections.

2. Connect the X_2 ends of two low-voltage secondary windings.

   Figure 21-4 illustrates two secondary coils with the X_2 coil ends connected. The voltage across the open ends should be 1.73 × 120 = 208 volts. However, if the leads on one transformer are reversed, the voltage across the open ends will be 120 volts.

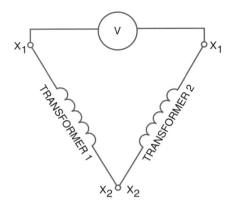

Figure 21-4 Two transformers correctly connected.

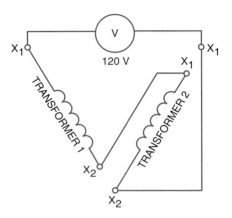

Figure 21-5 Two transformers incorrectly connected.

Figure 21-5 illustrates two transformers connected incorrectly. The voltage across the open ends is only 120 volts. If the leads of transformer 2 are reversed, the connections will be correct and the voltage across the open ends will be 208 volts.

3. Connect the X_2 lead of the low-voltage secondary winding of transformer 3 with the X_2 leads of the other two transformers.

The proper wye connection of the low-voltage secondary windings of the three single-phase transformers is shown in Figure 21-6. The voltage across each pair of open ends should be $1.73 \times 120 = 208$ volts. If the voltage across the open ends is correct, then the line leads feeding to the three-phase, 208-volt secondary system may be connected.

Figure 21-7 illustrates the secondary windings connected in wye with the line leads properly connected. Because each of the line wires is connected in series with one of the transformer windings, the current in each winding is equal to its respective line current.

Whenever single-phase transformers are connected in wye, the following current and voltage relationships are true.

* The line voltage is equal to $1.73 \times$ winding voltage.

* The line current and the winding current are equal.

The wye-wye connection scheme is satisfactory as long as the load on the secondary side is generally balanced. The wye-wye connection is unsatisfactory in situations where the secondary load becomes greatly unbalanced. An unbalanced load results in a serious unbalance in the three output voltages of the transformer bank.

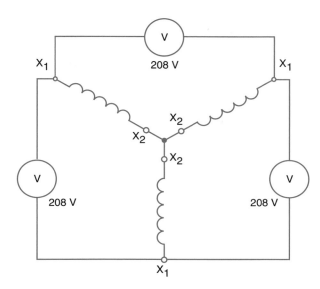

Figure 21-6 Three single-phase transformers properly connected in a wye arrangement.

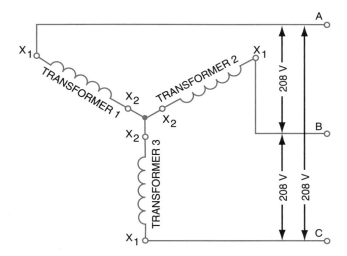

Figure 21-7 Three single-phase transformers properly connected to the line.

THREE-PHASE, FOUR-WIRE WYE CONNECTION

Voltage unbalancing in the secondary of the transformer bank can be nearly eliminated if a fourth wire (neutral wire) is used. This neutral wire connects between the source and the neutral point on the primary side of the transformer bank.

In the connection diagram (Figure 21-8), a three phase, four-wire system is used to feed the three-phase, high-voltage input to the transformer bank. The grounded neutral wire is connected to the common point where all three high-voltage primary winding ends

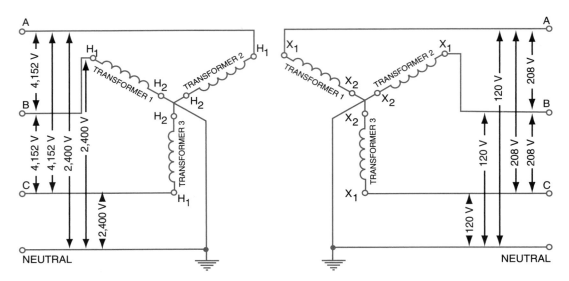

Figure 21-8 Wye-wye transformer bank with neutral connection.

or H_2 leads connect. The voltage from the neutral to any one of the three line wires is 2,400 volts. Each high-voltage winding is connected between the neutral and one of the three line loads. Therefore, 2,400 volts is applied to each of the three high-voltage primary windings. The voltage across the three line leads is $1.73 \times 2,400$ volts or 4,152 volts. The neutral wire maintains a relatively constant voltage across each of the high-voltage primary windings even though the load is unbalanced. Because the neutral wire is grounded, it helps protect the three high-voltage primary windings from lightning surges.

A three-phase, four-wire system also feeds from the low-voltage secondary side of the transformer bank to the load. Each low-voltage secondary winding is connected between the secondary grounded neutral and one of the three line leads. As on the primary side, the grounded neutral helps protect the low-voltage secondary windings from lightning surges.

The voltage output of each secondary winding is 120 volts. The voltage between the neutral and any one of the three line leads on the secondary side is 120 volts, as shown in Figure 21-8. The voltage across the three line leads is $1.73 \times 120 = 208$ volts. Thus, by using a three-phase, four-wire secondary, two voltages are available for different types of loads: 208 volts, three phase, for industrial power loads such as three-phase motors, and 120 volts, single phase, for lighting loads.

Many single-phase transformers are designed so that the low-voltage side consists of two 120-volt windings. These two windings can be connected in series for 240 volts or in parallel for 120 volts.

Figure 21-9 shows three single-phase transformers connected as a wye-wye bank. Each transformer has two 120-volt, low-voltage windings. For each single-phase

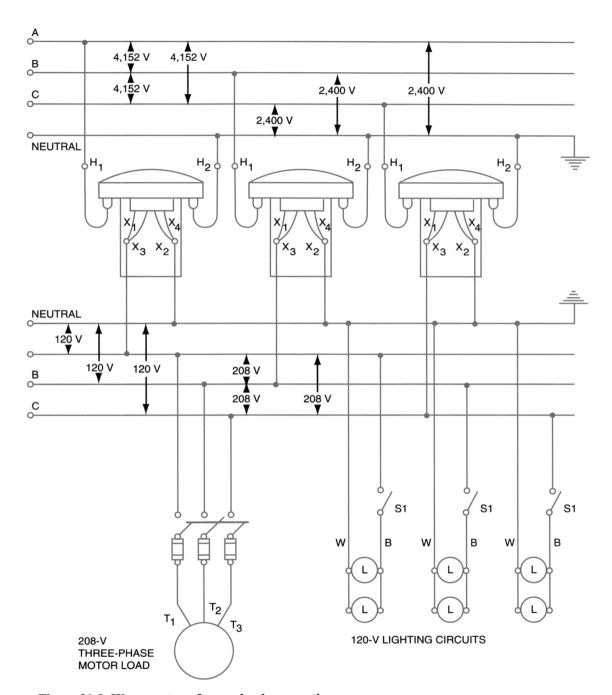

Figure 21-9 Wye-wye transformer bank connections.

transformer, the low-voltage coils are connected in parallel to give a voltage output of 120 volts. Note in Figure 21-9 that the secondary output windings of the three transformers are connected in wye. This three-phase, four-wire secondary system provides two different types of service:

- three-phase, 208-volt service for motor loads
- single-phase, 120-volt service for lighting loads

The 120/208-volt wye system is commonly used in schools, stores, and offices. Another popular system for large installations is the 480/277-volt wye system. Some applications of this system include

- motors connected to 480 volts (phase to phase)
- fluorescent lighting fixtures connected to 277 volts (phase to neutral)
- 120-volt outlets, incandescent lamps, and appliances connected to 120-volt circuits supplied from single-phase, 480/120/240-volt transformers or three-phase, 480/208Y/120-volt transformers. These separate transformers are connected to the 480-volt feeders for the primary source.

Three single-phase transformers of the same kilovolt-ampere capacity are used in most wye-wye connected transformer banks. The total kilovolt-ampere capacity of a wye-wye connected bank is found by adding the individual kVA ratings of the transformer. If each transformer is rated at 25 kVA, then the total kVA is 25 + 25 + 25 = 75 kVA.

If one transformer becomes defective, it must be replaced before the transformer bank can be reenergized. A wye-wye connected transformer bank cannot be reconnected in an emergency situation using only two single-phase transformers, such as in the open-delta system.

SUMMARY

Single-phase transformers can be connected in a wye or star pattern to achieve the desired three-phase and single-phase voltages required by many commercial customers. By connecting the single phases in a wye pattern, the line voltages can be increased by a factor of 1.73 times the coil voltage. This increased level of voltage is often desirable to reduce the line current drawn by a load. By increasing the voltage, the current will be less for a specific wattage load. Be sure to check all transformer polarities and check the final connections for a solid ground if using the three-phase, four-wire wye-connected system.

ACHIEVEMENT REVIEW

1. Draw a connection diagram for three wye-wye connected single-phase transformers. This transformer bank is used to step down 2,400/4,152 volts on a three-phase, four-wire primary to 120/208 volts on a three-phase, four-wire secondary.

Each transformer is rated at 20 kVA, with 2,400 volts on the high-voltage winding and 120 volts on the low-voltage winding. Mark leads H_1, X_1, and so forth; show all voltages.

2. What is the total kVA capacity of the wye-wye transformer bank in question 1?

    _____

    _____

    _____

3. A grounded neutral wire is used with a wye-wye connected transformer bank for what purpose?

    _____

    _____

    _____

4. The three-phase, four-wire secondary output of a wye-connected transformer bank can be used for what two types of load?

    a. _____

    _____

    b. _____

    _____

5. List the steps that may be used in connecting three single-phase transformers in wye.

   a. _____

   _____

   _____

   b. _____

   _____

   _____

   c. _____

   _____

   _____

6. When single-phase transformers are connected in a three-phase wye,

   a. what is the line current compared with the phase-winding current?

   _____

   _____

   b. what is the line voltage compared with the phase-winding voltage?

   _____

   _____

22

WYE AND DELTA CONNECTIONS OF SINGLE-PHASE TRANSFORMERS

OBJECTIVES

After studying this unit, the student should be able to

- diagram the connection of three single-phase transformers to form a delta-wye transformer bank.
- describe how a delta-wye transformer bank is used to step down voltages.
- describe how a delta-wye transformer bank is used to step up voltages.
- diagram the connection of three single-phase transformers to form a wye-delta transformer bank.
- diagram the connections of two single-phase transformers connected open-wye–open-delta.
- describe how a wye-delta transformer bank is used to step down voltages.
- list advantages and disadvantages of a single three-phase transformer as compared to three single-phase transformers.

Five commonly used methods of connecting single-phase transformers to form three-phase transformer banks are delta-delta, open-delta, wye-wye, delta-wye, and wye-delta connections. The delta-delta, open-delta, and wye-wye connections are described in previous units.

This unit covers the delta-wye and wye-delta connections by explaining the current and voltage relationships for each of these methods of three-phase transformation, and providing examples of several applications for each connection method.

STEP-DOWN APPLICATION FOR DELTA-WYE TRANSFORMER BANK

Assume that electrical energy must be transformed from a 2,400-volt, three-phase, three-wire input to a 120/208-volt, three-phase, four-wire output. Each of the three single-phase transformers is rated at 20 kVA, with 2,400 volts on the high-voltage windings and 120 or 240 volts on the low-voltage windings.

The primary windings of the three single-phase transformers are delta connected. The line voltage of the three-phase, three-wire primary input is 2,400 volts. Remember that the line voltage and the coil voltage are the same in a delta connection. As a result, the voltage across each of the primary coil windings is also 2,400 volts.

Figure 22-1 illustrates the connections for the delta-wye transformer bank in this example. Each transformer has two 120-volt, low-voltage windings that are connected in parallel to give a voltage output of 120 volts for each single-phase transformer. The secondary connections show that the output windings of the three transformers are connected in wye. Two types of service are available as a result of the three-phase, four-wire secondary system:

- three-phase, 208-volt service for motor loads
- single-phase, 120-volt service for lighting loads

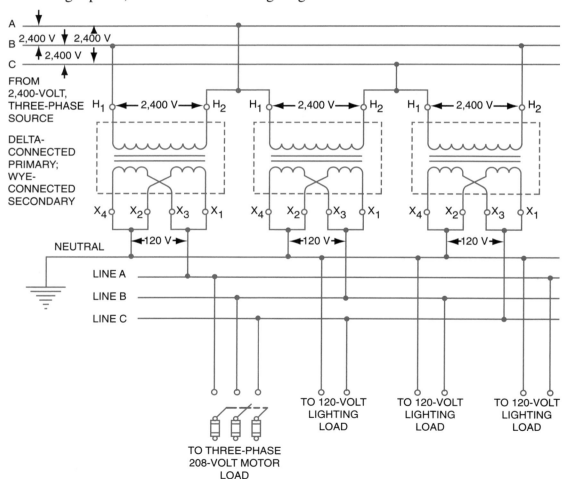

Figure 22-1 Delta-wye transformer bank.

The primary or input side of the bank in Figure 22-1 is delta connected. Therefore, for the primary side of the delta bank, the following are true:

- The line voltage and the coil voltage are the same.
- The line current is equal to $1.73 \times$ coil current.

The secondary or output side of this transformer bank is wye connected. For the secondary, the following are true:

- The line voltage is equal to $1.73 \times$ coil voltage.
- The line current and the coil current are equal.

The three single-phase transformers used in a delta-wye connection have the same kVA capacity. The transformers in this example are each rated at 20 kVA. The total kVA capacity of a delta-wye transformer bank is determined by adding the three kVA ratings. Because each transformer is rated at 20 kVA, the total delta-wye bank capacity is 60 kVA.

If one transformer becomes defective, it must be replaced before the bank can be reenergized. In an emergency situation, a delta-wye connected transformer bank cannot be reconnected using only two transformers.

In the delta-wye connection illustrated in Figure 22-1, the three single-phase transformers are connected to obtain additive polarity. However, the transformers that are used in an installation may have either additive or subtractive polarity. The polarity of each transformer must be checked. Then, if the basic rules for making delta connections and wye connections are followed, the electrician should have no difficulty in making standard three-phase transformer bank connections.

STEP-UP APPLICATION FOR DELTA-WYE TRANSFORMER BANK

The delta-wye transformer bank is well adapted for stepping up voltages. The input voltage is stepped up by the transformer ratio and is then increased further by the voltage relationship for a wye connection: line voltage = $1.73 \times$ coil voltages. In addition, the insulation requirements for the secondaries are reduced. This is an important advantage when very high voltages are used on the secondary side.

A delta-wye transformer bank used to step up the voltage at a generating station is illustrated in Figure 22-2. The high-voltage output is connected to three-phase transmission lines. These transmission lines deliver the electrical energy to municipal and industrial consumers who may be miles away from the generating station.

Transformer ratio refers to the actual ratio of primary to secondary voltage. *Transformation ratio* refers to the ratio of primary line voltage to secondary line voltage. In transformers that have different primary and secondary patterns, this distinction is important. In delta-wye, the transformer ratio may be 1 to 5, but the transformation ratio is 1 to 8.65.

As shown in Figure 22-2, the AC generators deliver energy to the generating station bus bars at a three-phase potential of 13,800 volts. The primary windings of the three single-phase transformers are each rated at 13,800 volts. These primary windings are connected in delta to the generating station bus bars; therefore, each primary coil winding has 13,800 volts applied to it. The transformers have a step-up ratio of 1 to 5. As a result, the voltage output of the secondary of each single-phase transformer is $5 \times 13,800 = 69,000$ volts. Figure 22-2 shows that the three secondary windings are connected in wye. Each high-voltage secondary winding is connected between the secondary neutral and one of the three line leads. The voltage between the neutral and any one of the three line leads is the same as the secondary coil voltage, or 69,000 volts. The voltage across the

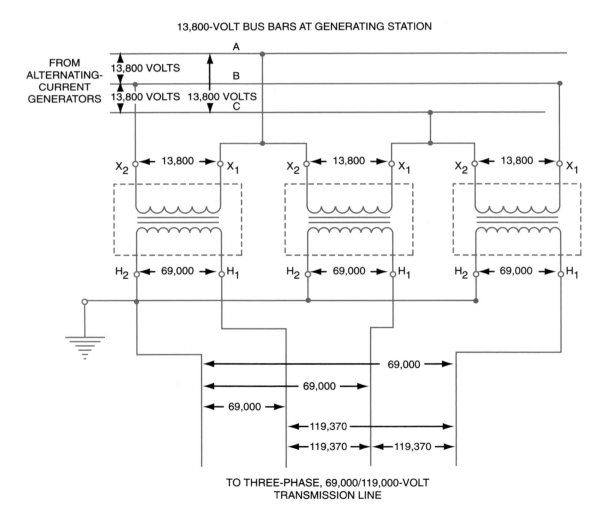

Figure 22-2 Delta-wye transformer bank.

three line leads is $1.73 \times 69{,}000 = 119{,}370$ volts. The grounded neutral wire on the high-voltage secondary output must be used to obtain balanced three-phase voltages even though the load current may be unbalanced. Not only is this neutral wire grounded at the transformer bank, it is also grounded at periodic intervals on the transmission line. As a result, it protects the three high-voltage secondary windings of the single-phase transformers from possible damage due to lightning surges.

WYE-DELTA TRANSFORMER BANK

A transformer bank connected in wye-delta is the type most often used to step down relatively high transmission line voltages (60,900 volts or more) at the consumer's location. Two reasons for selecting this type of transformer bank are that the three-phase voltage is decreased by the transformer ratio multiplied by the factor 1.73, and the insulation requirements for the high-voltage primary windings are reduced.

As an example, assume that it is necessary to step down a three-phase 60,900-volt input to a three-phase, 4,400-volt output (Figure 22-3). The primary windings are connected in wye to a three-phase, four-wire transmission line. The three-line voltages are

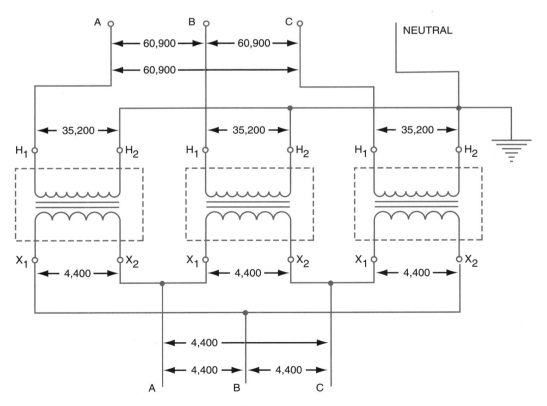

Figure 22-3 Wye-delta transformer bank.

60,900 volts each between phase conductors, and the voltage from each line wire to the grounded neutral is 35,200 volts ($\frac{60,900}{1.73}$).

Each of the three single-phase transformers is rated at 1,000 kVA, with 35,200 volts on the high-voltage side and 4,400 volts on the low-voltage side. The voltage ratio of each transformer is 8 to 1.

Figure 22-3 shows that the secondary windings are connected in delta, resulting in a line voltage of 4,400 volts on the three-phase, three-wire secondary system feeding to the load.

The total kVA capacity of a wye-delta transformer bank is determined by adding the kVA rating of each single-phase transformer in the bank. For the bank in Figure 22-3, the total kVA capacity is equal to 1,000 + 1,000 + 1,000 = 3,000 kVA.

OPEN-WYE–OPEN DELTA TRANSFORMER

A variation of the wye-delta transformer pattern is the open-wye–open-delta transformer connection. Another name for this connection is the wye-delta with one transformer missing. This is sometimes used to deliver small amounts of three-phase power to a secondary load and deliver larger amounts of single-phase power to the secondary load.

The connections are made to a three-phase system, but with only two single-phase transformers in the bank. The primary winding is connected as in Figure 22-4 to two of the three phase conductors and the common point is connected to the grounded neutral point. The voltage on the primary is fed from two of the phases of a three-phase system so the waveforms are 120° out of phase. Because the primary windings are really across the line voltage and the two primary coils are essentially equal values and in series, the voltage splits equally between the two coils. In Figure 22-4 it would be 240 V per coil. If we use transformers rated at 2400 VA, transformer 1 will have 10 amps of primary current at 240 V. Likewise, transformer 2 will have 10 A of primary current. The currents add at the neutral to become 17.3 amps, or 1.73 times the phase current, flowing in the neutral conductor. This large current in the neutral can create problems. The conductor must be sized for the higher current (1.73 × the phase current) and it tends to cause imbalance in the phase voltages.

The secondary coils are connected in the open-delta pattern with one transformer's center tap connection being grounded. We now have a three-phase secondary with three-phase line voltage and single-phase voltage available. This a true three-phase–four-wire secondary.

Calculations for the open-delta are the same as for other open-delta connections. The line voltage of the three-phase system is the coil voltage (in Figure 22-4 it is 240 V). The line current is the same as the phase current (in Figure 22-4, it is 10 A). Three-phase power is E(line) × I (line) × 1.73 = 3 phase power. For example, 240 × 10 × 1.73 = 4152 VA.

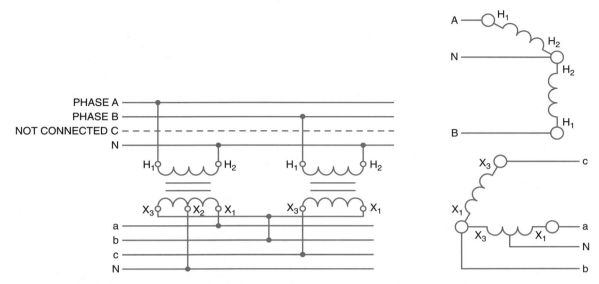

Figure 22-4 Open-wye–open-delta transformer pattern wiring diagram and schematic connections.

The primary calculations are not the same as for a completed wye. The primary uses phase voltage as the voltage of the source; the current from the lines is the same as the coil current. The formula then is voltage (coil) × current (line) × 1.73 = 3-phase power. For example, in Figure 22-4, 240 V × 10 A × 1.73 = 4152 VA.

You will notice that the three-phase capacity of the bank is less than the two transformer KVAs added together. For example, if two 2400-VA transformers were added together arithmetically, it would be 4800 VA. However, the transformer bank's total VA of two 2400-VA transformers' capacity is only 4152 VA. Because one transformer is missing, the total capacity of each transformer is not used and the total capacity of the bank is 86.5 percent of the arithmetic sum.

This pattern is used to supply a significant single-phase load and a light three-phase load. In that application the two transformers are not the same KVA rating. The center tapped transformer would be sized to deliver the three-phase load, plus the additional KVA of the single-phase load.

THREE-PHASE TRANSFORMERS

Voltages on three-phase systems may be transformed using three-phase transformers. The core of a three-phase transformer is made with three legs. A primary and a secondary winding of one phase are placed on each of the three legs. These transformers may be connected in delta-delta, wye-wye, delta-wye, or wye-delta. The

Figure 22-5 A three-phase transformer (assembled core and coils for a 500-kVA, 60-Hz, 13,800- to 2,400-V transformer).

Figure 22-6 34-kV power transformer at a generating station.

connections are made inside the transformer case. For delta-delta connections, three high-voltage and three low-voltage leads are brought out. Four leads are brought out when any wye-connected windings are used. This fourth lead is necessary for the neutral wire connection.

The three-phase transformer occupies less space than three single-phase transformers because the windings can be placed on one core in the three-phase transformer case (Figure 22-5). The efficiency of a three-phase transformer is higher than the overall efficiency of three single-phase transformers connected in a three-phase bank.

However, there is one disadvantage to the use of a three-phase transformer. If one of the phase windings becomes defective, the entire three-phase unit must be taken out of service. If a single-phase transformer in a three-phase bank becomes defective, it can be replaced quickly. The resultant power interruption is brief. For this reason, many transformer installations consist of banks of three single-phase transformers. Figure 22-6 shows a three-phase single unit transformer in operation.

SUMMARY

Transformers need not be connected in the same pattern on the primary and the secondary. Depending on the desired voltage level and level of step-up (increase) or step-down (decrease), the patterns may change. The two most popular patterns are the wye and delta.

To get the greatest step-up, the transformation ratio is best if the primary is connected delta and the secondary is connected wye. Likewise, to get the largest decrease in voltage, the ratio of transformation is the greatest if the primary is connected wye and the secondary is connected delta. Remember that the current ratios are the inverse of the voltage ratios.

ACHIEVEMENT REVIEW

1. Diagram the connections for three single-phase transformers connected in delta-wye to step down 2,400 volts, three phase, three wire to a 120/208-volt, three-phase, four-wire service. Three single-phase transformers are to be used. Each transformer is rated at 25 kVA, with 2,400 volts on the high-voltage side and 120 volts on the low-voltage side. Mark leads H_1, X_1, and so forth. Show all voltages.

2. What is the total kVA capacity of the delta-wye transformer bank in question 1?

   _____

3. What are two applications of a three-phase, delta-wye transformer bank?

   a. _____

   _____

   b. _____

   _____

4. What is one practical application of a three-phase, wye-delta transformer bank?

   _____

   _____

   _____

5. What is the main advantage of using the open-wye–open-delta transformer pattern?

   _____

   _____

   _____

6. Diagram the connections for three single-phase transformers connected in wye-delta to step down a three-phase input of 33,000 volts to a three-phase output of 4,800 volts. Mark leads H_1, X_1, and so forth. Show all voltages.

7. What is one advantage to using a three-phase transformer in place of three single-phase transformers?

   _____

   _____

   _____

8. What is one disadvantage to using a three-phase transformer in place of three single-phase transformers?

   _____

   _____

   _____

   _____

9. Insert the word or phrase that completes each of the following statements.
   a. A wye-delta transformer bank has _____-connected primary windings and_____-connected secondary windings.
   b. A delta-wye transformer bank has _____-connected primary windings and_____-connected secondary windings.
   c. A wye-delta transformer bank is used to _____ extremely high three-phase voltages.
   d. A three-phase transformer takes _____ space than a transformer bank of the same kVA capacity consisting of three single-phase transformers.
   e. A three-phase transformer has a _____ percent efficiency than a transformer bank consisting of three single-phase transformers.

10. List five common three-phase connections used to connect transformer banks consisting of either two or three single-phase transformers.

    a. _____

    b. _____

    c. _____

    d. _____

    e. _____

11. What is the purpose of the grounded neutral on a three-phase, four-wire system?

    _____

    _____

    _____

Select the correct answer for each of the following statements and place the corresponding letter in the space provided.

12. A delta-wye, four-wire secondary gives   _____
    a. 120-volt, single-phase and 208-volt, three-phase output.
    b. 208-volt, single-phase and 120-volt, three-phase output.
    c. 208-volt, three-phase output.
    d. 120-volt, three-phase output.

13. Most three-phase systems use three single-phase   _____
    transformers connected in a bank because
    a. one transformer can be readily replaced if it becomes defective.
    b. better regulation is maintained.
    c. they are easier to cool.
    d. this method of connection is the most efficient.

14. Transformer capacities may be increased by   _____
    a. connecting them in series.
    b. pumping the oil.
    c. cooling the oil with fans.
    d. reducing the load.

15. A step-down, delta-wye transformer connection is   _____
    commonly used for
    a. motor and lighting loads.
    b. distribution of electrical energy.
    c. motor loads.
    d. lighting loads.

16. In a delta connection,   _____
    a. the line voltage and coil voltage are equal.
    b. the line current is equal to 1.73 times the coil current.
    c. the coils are connected in closed series.
    d. all of the above are true

17. In a wye connection,   _____
    a. the line current is equal to 1.73 times the coil current.
    b. the line voltage is equal to 1.73 times the coil voltage.
    c. the line voltage and coil voltage are equal.
    d. none of the above is true

23

INSTRUMENT TRANSFORMERS

OBJECTIVES

After studying this unit, the student should be able to

- explain the operation of an instrument potential transformer.
- explain the operation of an instrument current transformer.
- diagram the connections for a potential transformer and a current transformer in a single-phase circuit.
- state how the following quantities are determined for a single-phase circuit containing instrument transformers: primary current, primary voltage, primary power, apparent power, and power factor.
- describe the connection of instrument transformers in a three-phase, three-wire circuit.
- describe the connection of instrument transformers to a three-phase, four-wire system.

Instrument transformers are used to measure and control AC circuits. Direct measurement of high voltage or heavy currents involves large and expensive instruments, relays, and other circuit components of many designs. Using instrument transformers, however, makes it possible to use relatively small and inexpensive instruments and control devices of standardized designs. Instrument transformers also protect the operator, the measuring devices, and the control equipment from the dangers of high voltage. The use of instrument transformers results in increased safety, accuracy, and convenience.

There are two distinct classes of instrument transformers: the instrument potential transformer and the instrument current transformer. (The word "instrument" is usually omitted for brevity.)

POTENTIAL TRANSFORMERS

The potential transformer operates on the same principle as a power or distribution transformer. The main difference is that the capacity of a potential transformer is small compared to that of power transformers. Potential transformers have ratings from 100 to 500 volt-amperes (VA). The low-voltage side is usually wound for 115 volts or 120 volts. The load on the low-voltage side usually consists of the potential coils of various instruments, but may also include the potential coils of relays and other control equipment. In general, the load is relatively light, and it is not necessary to have potential transformers with a capacity greater than 100 to 500 volt-amperes.

The high-voltage primary winding of a potential transformer has the same voltage rating as the primary circuit. When it is necessary to measure the voltage of a 4,600-volt, single-phase line, the primary of the potential transformer would be rated at 4,600 volts and the low-voltage secondary would be rated at 115 volts. The ratio between the primary and secondary windings is

$$\frac{4,600}{115} = \frac{40}{1}$$

A voltmeter connected across the secondary of the potential transformer indicates a value of 115 volts. To determine the actual voltage on the high-voltage circuit, the instrument reading of 115 volts must be multiplied by 40 ($115 \times 40 = 4,600$ volts). In most cases, the voltmeter is calibrated to indicate the actual value of voltage on the primary side. As a result, the operator is not required to apply the multiplier to the instrument reading, and the possibility of errors is reduced.

Figure 23-1 illustrates the connections for a potential transformer with a 4,600-volt primary input and a 115-volt output to the voltmeter. This potential transformer has subtractive polarity. (All instrument potential transformers now manufactured have subtractive polarity.) One of the secondary leads of the transformer in Figure 23-1 is grounded to eliminate high-voltage hazards.

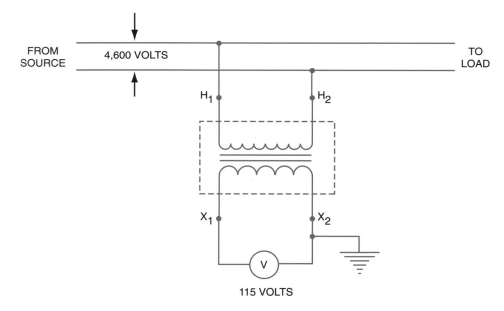

Figure 23-1 Connections for a potential transformer.

Potential transformers have highly accurate ratios between the primary and secondary voltage values; generally the error is less than 0.5 percent. Power transformers are not designed for highly accurate voltage transformation.

CURRENT TRANSFORMERS

Current transformers are used so that ammeters and the current coils of other instruments and relays need not be connected directly to high-current lines. In other words, these instruments and relays are insulated from high currents. Current transformers also step down the current to a known ratio. The use of current transformers means that relatively small and accurate instruments, relays, and control devices of standardized design can be used in circuits.

The current transformer has separate primary and secondary windings. The primary winding, which may consist of a few turns of heavy wire wound on a laminated iron core, is connected in series with one of the line wires. The secondary winding consists of a greater number of turns of a smaller size of wire. The primary and secondary windings are wound on the same core.

The current rating of the primary winding of a current transformer is determined by the maximum value of the load current. The secondary winding is rated at 5 amperes regardless of the current rating of the primary windings.

For example, assume that the current rating of the primary winding of a current transformer is 100 amperes. The primary winding has three turns, and the secondary winding has 60 turns. The secondary winding has the standard current rating of 5 amperes; therefore, the ratio between the primary and secondary currents is 100/5 or 20 to 1. The primary current is 20 times greater than the secondary current. Because the secondary winding has 60 turns and the primary winding has 3 turns, the secondary winding has 20 times as many turns as the primary winding. For a current transformer, then, the ratio of primary to secondary currents is inversely proportional to the ratio of primary to secondary turns.

In Figure 23-2, a current transformer is used to step down current in a 4,600-volt, single-phase circuit. The current transformer is rated at 100 to 5 amperes, and the ratio of current step-down is 20 to 1. In other words, there are 20 amperes in the primary winding for each ampere in the secondary winding. If the ammeter at the secondary indicates 4 amperes, the actual current in the primary is 20 times this value, or 80 amperes.

The current transformer in Figure 23-2 has polarity markings in that the two high-voltage primary leads are marked H_1 and H_2, and the secondary leads are marked X_1 and X_2. When H_1 is instantaneously positive, X_1 is positive at the same moment. Some current transformer manufacturers mark only the H_1 and X_1 leads or use polarity marks. Polarity marks can be any similar marks that are placed on the H_1 lead and the X_1 lead. Common marks are Xs, dots, or squares. When connecting current transformers in circuits, the

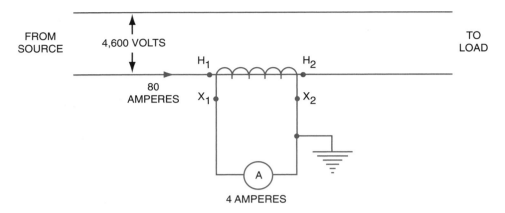

Figure 23-2 A current transformer used with an ammeter.

H_1 lead is connected to the line lead feeding from the source, while the H_2 lead is connected to the line lead feeding to the load.

The secondary leads are connected directly to the ammeter. Note that one of the secondary leads is grounded as a safety precaution to eliminate high-voltage hazards.

Caution: The secondary circuit of a transformer should never be opened when there is current in the primary winding. If the secondary circuit is opened when there is current in the primary winding, then the entire primary current is an exciting current that induces a high voltage in the secondary winding. This voltage can be high enough to endanger human life.

Individuals working with current transformers must check that the secondary winding circuit path is closed. At times, it may be necessary to disconnect the secondary instrument circuit when there is current in the primary winding. For example, the metering circuit may require rewiring or other repairs may be needed. To protect a worker, a small short-circuiting switch is connected into the circuit at the secondary terminals of the current transformer. This switch is closed when the instrument circuit must be disconnected for repairs or rewiring.

Current transformers have very accurate ratios between the primary and secondary current values: the error of most modern current transformers is less than 0.5 percent.

When the primary winding has a large current rating, it may consist of a straight conductor passing through the center of a hollow metal core. The secondary winding is wound on the core. This assembly is called a bar-type current transformer. The name is derived from the construction of the primary, which actually is a straight copper bus bar. All standard current transformers with ratings of 1,000 amperes or more are bar-type transformers. Some current transformers with lower ratings may also be of the bar type. Figure 23-3 shows a bar-type current transformer.

In window-type current transformers, the primary passes through an opening in the transformer windings. The primary of the transformer is the line conductor, much like the bar forms the primary in the bar-type current transformer. The shape of the transformer resembles a donut. Therefore, this type of current transformer (CT) is often referred to as a "donut."

Figure 23-3 Bar-type current transformer.

Figure 23-4 shows a clamp-on ammeter that uses the concept of a window-type current transformer. By opening the clamp and then closing it around the current-carrying conductor, the current in the conductor is measured on the meter.

INSTRUMENT TRANSFORMERS IN A SINGLE-PHASE CIRCUIT

Figure 23-5 illustrates an instrument load connected through instrument transformers to a single-phase, high-voltage line. The instruments include a voltmeter, an ammeter, and a wattmeter. The potential transformer is rated at 4,600 to 115 volts; the current transformer is rated at 50 to 5 amperes. The potential coils of the voltmeter and the wattmeter are connected in parallel across the low-voltage output of the potential transformer. Therefore, the voltage across the potential coils of each of these instruments is the same. The current coils of the ammeter and the wattmeter are connected in series across the secondary output of the current transformer. As a result, the current in the

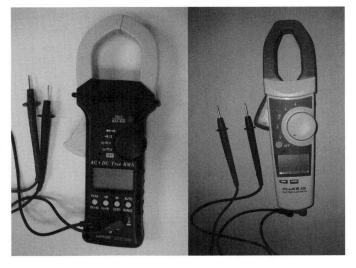

Figure 23-4 Clamp-on style ammeters/multimeters.

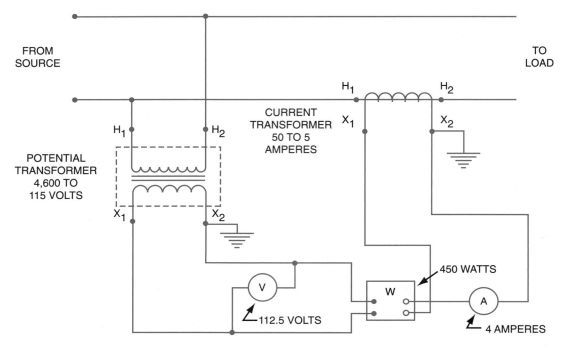

Figure 23-5 Single-phase metering connections.

current coils of both instruments is the same. Note that the secondary of each instrument transformer is grounded to provide protection from high-voltage hazards, as provided in *Article 250* of the *National Electrical Code*®.

The voltmeter in Figure 23-5 reads 112.5 volts, the ammeter reads 4 amperes, and the wattmeter reads 450 watts. To find the primary voltage, primary current, primary power, apparent power in the primary circuit, and the power factor, the following procedures are used:

Primary Voltage

$$\text{Voltmeter multiplier} = 4{,}600/115 = 40$$

$$\text{Primary volts} = 112.5 \times 40$$

$$= 4{,}500 \text{ volts}$$

Primary Current

$$\text{Ammeter multiplier} = 50/5 = 10$$

$$\text{Primary amperes} = 4 \times 10$$

$$= 40 \text{ amperes}$$

Primary Power

Wattmeter multiplier = Voltmeter multiplier × ammeter multiplier

Wattmeter multiplier = 40 × 10

= 400

Primary watts = 450 × 400

= 180,000 watts or 180 kilowatts

Apparent Power

The apparent power of the primary circuit is found by multiplying the primary voltage and current values.

Apparent power (volt-amperes) = volts × amperes

Volt-amperes = 4,500 × 40

$$= 180,000 \text{ volt-amps} = \frac{180,000}{1,000} = 180 \text{ KVA}$$

Power Factor

$$\text{Power factor} = \frac{\text{Power in kilowatts}}{\text{Apparent power in kilovolt-amperes}}$$

= 180/180

= 1.00 or 100 percent

In DC instruments needed to monitor large values of DC, another type of system is needed. Because it is not possible to use standard transformers to change DC levels, instrument shunts are used to develop the smaller values needed for DC monitoring equipment. In effect, these DC shunts are high-power, low-resistance resistors that are placed in line with the DC power. They typically have connection points that are used to connect to meters. The resistance of the shunt is designed so that when full line current flows through the resistive material, it will develop 50 millivolts across the resistance and therefore the meter movement. (See the DC meter in Figure 23-6.)

INSTRUMENT TRANSFORMERS ON THREE-PHASE SYSTEMS

Three-Phase, Three-Wire System

On a three-phase, three-wire system, two potential transformers of the same rating and two current transformers of the same rating are necessary. It is common practice in

Figure 23-6 Panel-mounted meters use transformers to monitor large values.

three-phase metering to interconnect the secondary circuits. That is, the connections are made so that one wire or device conducts the combined currents of two transformers in different phases.

The low-voltage instrument connections for a three-phase, three-wire system are shown in Figure 23-7. Note that the two potential transformers are connected in open delta to the 4,600-volt, three-phase line. This results in three secondary-voltage values of 115 volts each. The two current transformers are connected so that the primary of one transformer is in series with line A and the primary winding of the second transformer is in series with line C.

Note that three ammeters are used in the low-voltage secondary circuit. This wiring system is satisfactory on a three-phase, three-wire system, and all three ammeters give accurate readings. Other instruments that can be used in this circuit include a three-phase wattmeter, a three-phase watt-hour meter, and a three-phase power factor meter. When three-phase instruments are connected in the secondary circuits, these instruments must be connected correctly so that the proper phase relationships are maintained. If this precaution is not observed, the instrument readings will be incorrect. In checking the connections for this three-phase, three-wire metering system, note that the interconnected potential and current secondaries are both grounded to provide protection from high-voltage hazards.

Three-Phase, Four-Wire System

Figure 23-8 illustrates the secondary metering connections for a 2,400/4,152-volt, three-phase, four-wire system. The three potential transformers are connected in wye to give a three-phase output of three secondary voltages of 120 volts to neutral. Three

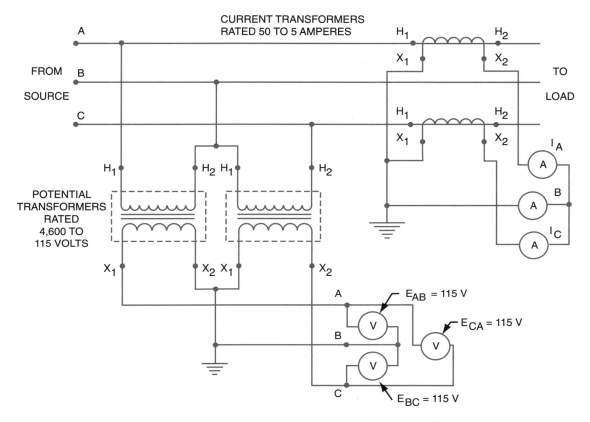

Figure 23-7 Metering connections for three-phase, three-wire system.

50- to 5-ampere current transformers are used in the three line conductors. Three ammeters are used in the interconnected secondary circuit. Both the interconnected potential and the current secondaries are grounded to protect against possible high-voltage hazards.

SUMMARY

Instrument transformers are specifically designed to transform voltage and current in very precise ratios. Potential transformers are used to transform high voltages to usable values of 115 or 120 volts for use by standard instruments. Current transformers are used to transform large values of AC down to a 5-amp level so that it can be used by standard instruments. DC levels are typically reduced to a usable level through the use of shunts. The shunt has a primary-load current rating and the meter is then connected across the shunt. The meter is designed to operate at 50 millivolts.

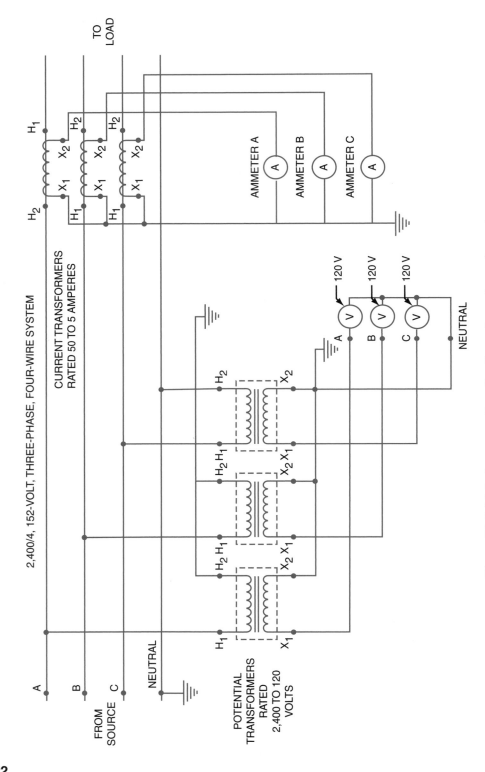

Figure 23-8 Metering connections for three-phase, four-wire system.

ACHIEVEMENT REVIEW

1. What are the two types of instrument transformers?

    a. _____   b. _____

2. Why must the secondary circuit of a current transformer be closed when there is current in the primary circuit? _____

    _____

    _____

3. A transformer is rated at 4,600/115 volts. A voltmeter connected across the secondary reads 112 volts. What is the primary voltage?   _____

4. A current transformer is rated at 150/5 amperes. An ammeter in the secondary circuit reads 3.5 amperes. What is the primary current?   _____

5. A 2,300/115-volt potential transformer and a 100/5-ampere current transformer are connected on a single-phase line. A voltmeter, an ammeter, and a wattmeter are connected in the secondaries of the instrument transformers. The voltmeter reads 110 volts, the ammeter reads 4 amperes, and the wattmeter reads 352 watts. Draw the connections for this circuit. Mark leads H_1, X_1, and so forth. Show all voltage, current, and wattage readings.

6. Complete a circuit using instrument transformers to measure voltage and amperage. Include terminal markings.

    FROM SOURCE   TO LOAD

7. What is the primary voltage of the single-phase circuit   _____
   in question 5?

8. What is the primary current in amperes of the single-phase   _____
   circuit in question 5?

9. What is the primary power in watts in the single-phase circuit   _____
   in question 5?

10. What is the power factor of the single-phase circuit   _____
    in question 5?

Select the correct answer for each of the following statements and place the corresponding letter in the space provided.

11. The secondary for a potential transformer is usually wound for   _____
    a. 10 volts.   c. 230 volts.
    b. 115 volts.   d. 500 volts.

12. Potential transformer secondaries are grounded to   _____
    a. stabilize meter readings.
    b. ensure readings with an accuracy of 0.5 percent.
    c. complete a system with the primaries.
    d. eliminate high-voltage hazards.

13. A transformer used to reduce current values to a size where   _____
    small meters can register them is a(n)
    a. autotransformer.   c. potential transformer.
    b. distribution transformer.   d. current transformer.

14. The primary of a large current transformer may consist of   _____
    a. many turns of fine wire.
    b. a few turns of fine wire.
    c. many turns of heavy wire.
    d. a straight-through conductor.

15. The standard ampere rating of the secondary of a current   _____
    transformer is
    a. 5 amperes.   c. 15 amperes.
    b. 50 amperes.   d. 150 amperes.

16. The secondary circuit of a current transformer should never   _____
    be opened when current is present in the primary because
    a. the meter will burn out.
    b. the meter will not operate.
    c. dangerous high voltage may develop.
    d. primary values may be read on the meter.

24

THREE-PHASE TRANSFORMERS

OBJECTIVES

After studying this unit, the student should be able to

- identify three-phase transformers.
- determine the lead identification of three-phase transformers.
- explain the efficiencies involved.
- determine the benefits and the detriments of three-phase transformers.

Three-phase transformer units are designed to be installed as a complete unit. Instead of installing three individual transformers and field connecting them into the desired pattern, a transformer (preassembled as a unit) is used. The transformer windings are assembled on a common core, and the appropriate leads are brought out. There are usually three high-voltage leads marked H_1–H_2–H_3. The secondary leads would be marked X_1–X_2–X_3. In three-phase transformers, the phase rotation or phase sequence between the primary and the secondary is critical.

All phase windings within the transformer normally have the same relative polarity. This means that if the transformer connection is a subtractive polarity, then the other phase connections would also be subtractive. However, the three-phase polarity depends on how the leads are brought out to the secondary terminals. Terminal markings alone do not indicate all the relationships between primary and secondary. Three-phase transformers should have a voltage vector diagram to show the angular phase displacement between the primary and the secondary and also the phase-sequence order. Figure 24-1 shows a voltage vector diagram for a three-phase transformer connected delta to delta. The American

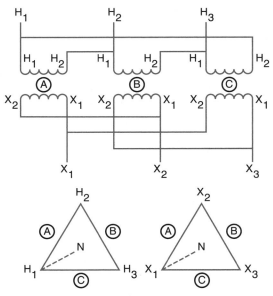

Figure 24-1 Internal connections determine 0° angular displacement.

National Standards Institute (ANSI) defines the angular displacement as the angle between H_1-N and X_1-N, where N is the neutral point in the vector diagram. In Figure 24-1, the displacement is 0°. This means that the relationship of primary phase sequence is H_1, H_2, H_3, and the secondary phase rotation is X_1, X_2, X_3. The same delta-to-delta configuration of a three-phase transformer may have an angular displacement of 180°. In this case, the vector diagram would appear as shown in Figure 24-2. The difference is in the internal connection of the secondary coils. These changes in secondary connections can be seen in the coil connection diagrams of Figures 24-1 and 24-2.

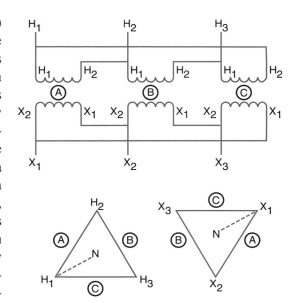

Figure 24-2 Internal connections to secondary winding determine 180° angular displacement.

Most three-phase, dry-type distribution transformers are connected as a primary delta connection. The secondary can be connected as a wye or a delta, and the leads brought out to the connection terminals. If the pattern is a delta-to-delta pattern, the displacement may be 0° or 180°. If the secondary pattern is a wye, the angular displacement is typically 30°. Figure 24-3 illustrates what a delta-wye pattern may look like. Figure 24-4 shows a three-phase transformer nameplate

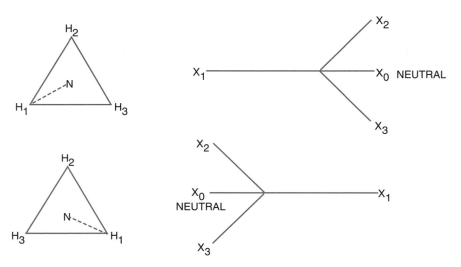

Figure 24-3 Δ-Y transformers are in group three transformers with 30° angular displacement.

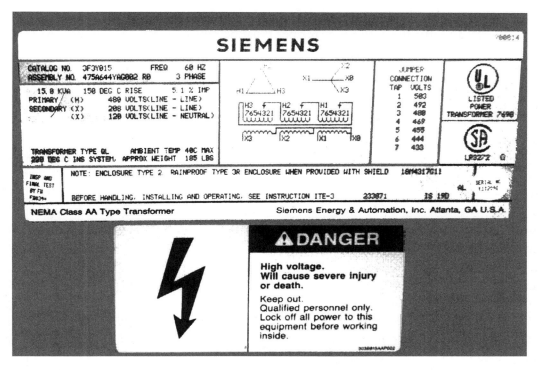

Figure 24-4 Transformer nameplates.

with the voltage vector diagrams and also the tap connection pattern for connecting primary voltages from 580 volts to 433 volts to obtain 208-volt, three-phase output.

PARALLELING THREE-PHASE TRANSFORMERS

Knowing the voltage vectors will allow you to properly parallel three-phase transformers for increased load capacity. It is only necessary to connect the similarly marked high- and low-voltage terminals if the transformers have the same voltage ratio, the same percent impedance, and the same angular displacement.

If the transformers are delta-wye or wye-delta, the angular displacement is 30°. When paralleling three-phase transformer units, only transformers with the same displacement should be connected. The only way to change the angular displacement is to reconnect the internal leads of the individual coils.

Advantages and Disadvantages

Because three-phase transformers are wound on the same core, the efficiency of transformation is higher, with less flux leakage than three single-phase units. Typically the cost is less for a three-phase unit compared to the same capacity system using three single-phase units. The disadvantage of a single unit housing a polyphase transformer is that if one coil fails, the entire transformer must be replaced, rather than just one transformer phase.

SUMMARY

Three-phase transformers are easier to install in transformer installations because all the internal connections are done. The same *NEC*® regulations apply for three single-phase or for one polyphase transformer installation. Care must be taken when connecting multiple three-phase transformers to supply a load. The connection pattern and the phase displacement must be considered. Three-phase transformers are more efficient in the transformation of power because of the common core and winding system. The disadvantage of the three-phase unit transformer is that if one phase fails, the whole transformer must be replaced.

ACHIEVEMENT REVIEW

1. In three-phase transformers, what is meant by 0° angular displacement?

     _____

     _____

     _____

     _____

2. List some of the conditions to be observed in paralleling three-phase transformers.

     _____

     _____

     _____

     _____

     _____

3. What are some of the advantages of three-phase tranformers, compared to three single-phase transformers? _____

     _____

     _____

     _____

     _____

     _____

4. What could be a disadvantage to connecting a three-phase transformer instead of three single-phase transformers? _____

    _____

    _____

    _____

5. How does the *NEC*® distinguish between three single-phase transformers and a single three-phase transformer? _____

    _____

    _____

    _____

    _____

6. Name two possible patterns that can be connected when using three-phase trans-formers._____

    _____

    _____

    _____

    _____

U • N • I • T

25

NATIONAL ELECTRICAL CODE®
REQUIREMENTS FOR
TRANSFORMER INSTALLATIONS

OBJECTIVES

After studying this unit, the student should be able to

- use the *National Electrical Code*® (*NEC*®) to determine the requirements and limitations of transformer installations.

Article 450 of the *NEC*® covers the minimum requirements of the installation of electrical wiring and equipment within public or private buildings and their premises.

TRANSFORMER LOCATION

The location of transformers is a prime *Code*® ruling. Most electrical codes and power companies state that transformers and transformer vaults must be readily accessible to qualified personnel for inspection and maintenance. The codes also contain specific sections covering oil-insulated, askarel-insulated, other dielectric fluid–insulated, and dry-type transformers, as well as transformer vaults. Dry-type transformers installed outdoors should have weatherproof enclosures.

TRANSFORMER OVERCURRENT PROTECTION

The *NEC*® provides information on the overcurrent protection required for transformers and transformer banks, as well as the maximum overcurrent protection allowed on a transformer primary.

Figure 25-1 illustrates where transformer primary protection is located.

For a transformer of 600 volts or less, overcurrent protection is permitted as seen in *Table 450.3B* of the *NEC*®.

Each transformer with a primary or secondary rated at 600 volts or less shall have overcurrent protection corresponding to the following conditions:

1. If there is protection of the primary conductors only and the primary current is less than 2 amps, the overcurrent protection must not exceed 300 percent of the primary

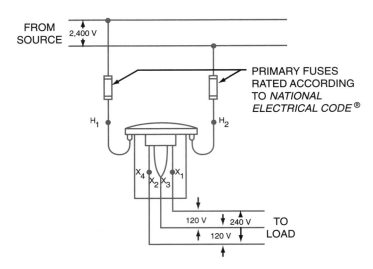

Figure 25-1 Transformer overcurrent protection.

current. If the primary current is between 2 and 9 amps, the overcurrent protection must not exceed 167 percent of the primary current. If the current is 9 amps or more, the overcurrent is to be 125 percent of the current or the next higher standard size of fuse or circuit breaker.

2. If there is primary and secondary overcurrent protection, then generally 250 percent can be used for the primary protection rating. (See Note 3, *Table 450.3(B)* in the *NEC*®.) The secondary will have 167 percent protection for less than 9 amps or 125 percent protection for 9 amps or more.

Each transformer with a primary or secondary rated at over 600 volts shall have overcurrent protection according to the following conditions:

1. In any location, if the primary is over 600 volts and the transformer impedance is not over 6 percent, use a primary fuse at 300 percent or a circuit breaker at 600 percent. If the transformer has an impedance over 6 percent but not more than 10 percent, the fuse is still 300 percent, but the circuit-breaker protection is reduced to 400 percent.

2. If secondary protection is required as when used for separately derived systems, use *Table 450.3(A)* of the *NEC*®, including the notes for reference.

The *Code*® requires that instrument potential transformers have primary fuses. Fuses of different sizes are required for operation above and below 600 volts.

SECONDARY CONNECTIONS BETWEEN TRANSFORMERS

The *Code*® *Article 450.6* defines a secondary tie as a circuit operating at 600 volts or less between phases. This circuit connects two power sources or power supply points such as the secondaries of two transformers.

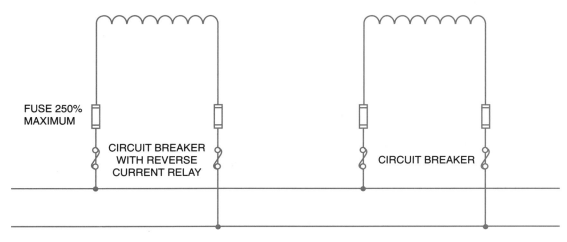

Figure 25-2 Transformer and feeder overcurrent protection.

When transformer secondaries are tied together to improve load performance, several provisions must be followed. If the transformers are connected as in Figure 25-2, *Article 450.6(B)* of the *Code*® is used. When the secondaries are tied, the transformers themselves must be protected by a secondary fuse rated at not more than 250 percent of the secondary rated current. In addition to this protection, there must be an automatic circuit breaker operated by a reverse current relay that is set to open the breaker at not more than the secondary current of the transformer. Then the transformer tie is sized and protected as described next.

A secondary tie circuit should have overcurrent protection at each end except in situations as described in *450.6(A)* of the *Code*®. A tie connection between two transformer secondaries is shown in Figure 25-3. Note that the tie conductor circuit has overcurrent protection at each end and no load taps are in the tie connections. The tap

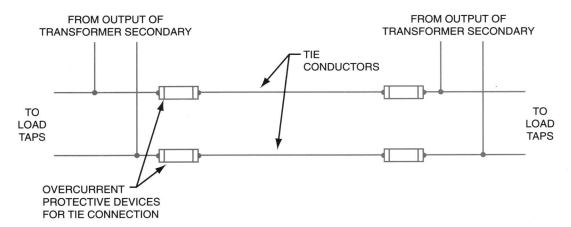

Figure 25-3 Tie connections between transformers.

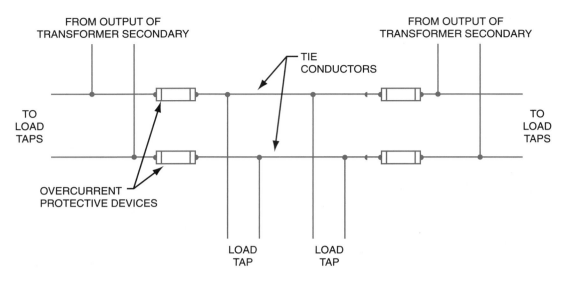

Figure 25-4 Tie connections between transformers with load taps.

conductor ampacity must be greater than or equal to 67 percent of the secondary of the largest transformer *(450.6(A)(1)* of the *NEC®*).

However, when load taps are made in the tie circuit between transformers, the minimum size of conductor required is regulated by the *Code®*. In this case, the current-carrying capacity shall not be less than a stipulated percentage (100 percent) of the rated secondary current of the largest capacity transformer connected to the secondary tie.

A tie connection (with load taps) between two transformer secondaries is shown in Figure 25-4. Because load taps are present, the size of the tie conductors must be increased to 100 percent of the secondary current of the largest transformer.

The following overload protective devices used for the tie connection with load taps are acceptable:

- limiting devices consisting of fusible-link cable connectors (limiters)
- automatic circuit breakers

If the voltage exceeds a value specified in the *Code®* Article *450.6(A)(5)*, the tie conductors must have a switch at each end of the tie circuit. When these switches are open, the limiters and tie conductors are deenergized. These switches shall be not less than the current rating of the tie conductors. Further, these switches shall be capable of opening their rated current.

PARALLEL OPERATION OF TRANSFORMERS

Transformers may be operated in parallel and protected as a unit if their electrical characteristics are similar. These electrical characteristics include the voltage ratio and the percentage impedance. When transformers have similar electrical characteristics, they will divide the load in proportion to their kVA rating. See *Article 450.7.*

GUARDING TRANSFORMERS

Appropriate provisions must be made to minimize the possibility of damage to transformers from external causes. This is particularly important if the transformers are located where they are exposed to mechanical damage.

Dry-type transformers must be provided with a noncombustible moisture-resistant case or enclosure that will provide reasonable protection against the accidental insertion of foreign objects. They also should have weatherproof enclosures when installed outdoors. The transformer installation must conform with the *Code*® provisions for guarding live parts.

The operating voltage of exposed live parts of transformers must be marked by warning signs or visible markings. These markings or signs are to be mounted in unobstructed positions on the equipment and structure.

GROUNDING

The *NEC*® requires that the metal cases and tanks of transformers be grounded. Further, all noncurrent-carrying metal parts of transformer installations and structures, including fences, are also to be grounded and bonded. This grounding must be done in the manner prescribed by the *Code*® to minimize any voltage hazard that may be caused by insulation failures or static conditions.

TRANSFORMER NAMEPLATE DATA

According to the *Code*® *Article 450.11,* each transformer shall be provided with a nameplate, and the nameplate must include the following information:

a. manufacturer's name

b. rated kVA capacity

c. frequency in hertz

d. primary and secondary voltages

e. amount of insulating liquid and type used

f. temperature class of insulation in dry-type transformers

g. impedance (25 kVA and higher)

h. required clearances for transformers with vented openings

The *NEC*® requires that the manufacturer's name (a) be on the nameplate to establish reliability and conformance to standard transformer designs. The full-load kVA (b) of the transformer is provided rather than the full-load wattage because the power factor of the load is not known by the manufacturer. It is more accurate to specify the kVA and allow the system designer to make the right application for the transformer. The frequency in hertz (c) is used to allow the proper magnetic effects to ensure accurate transformation of the rated values. The primary and secondary voltages (d) are given at full-load values, that is, when the transformer is operating at full kVA capacity. The amount of insulating liquid (e) is a necessary value when installing liquid-filled transformers. If the transformer should rupture, the liquid must be contained by a capture system. It is essential to know how much liquid must be contained in the event of a spill. In all dry-type transformers, the insulation class (f) must be listed, which will be an indication of the temperature that may be encountered while the transformer is operating. This value indicates how and where the dry transformer can be mounted and installed and its proximity to flammable materials. On transformers 25 kVA and larger, the impedance rating (g) must be listed. The impedance is an indication of how much opposition the transformer will offer to AC voltages. It is used to calculate the short circuit–current ratings that a transformer can deliver to the load when a short circuit occurs on the secondary side of the transformer. The last requirement on the nameplate is the clearance distance (h) required for transformers with vented openings for heat dissipation. Again, this aids the electrician when installing the transformer to maintain the proper distance from flammable materials.

DRY-TYPE TRANSFORMERS INSTALLED INDOORS

Dry-type transformers are used extensively for indoor installations. These transformers are cooled by air. They are not encased in the steel tanks required for oil-filled transformers. For protection, dry-type transformers are enclosed in sheet metal cases with openings to allow air to circulate.

The *Code*® specifies that dry-type transformers of a 112 1/2-kVA rating or less must have a separation of 12 in. from any combustible material; however, there are *Code*® conditions and exceptions.

Some transformers of more than a 112 1/2-kVA rating must be installed in a transformer room with fire-resistant construction or must be installed in a transformer vault. Transformers with Class 155 insulation (80°C temperature rise) need not be installed in a transformer vault provided they are separated from combustible material by the horizontal and vertical dimensions as specified in the *Code*®. Any dry-type transformer rated at more than 35,000 volts must be installed in a transformer vault.

ASKAREL-INSULATED TRANSFORMERS INSTALLED INDOORS

The windings of some transformers are cooled and insulated by a synthetic, nonflammable liquid called *askarel*. Askarel, when decomposed by an electric arc, produces only nonexplosive gases.

The *Code*® specifies that askarel-insulated transformers over 25 kVA must be furnished with a pressure-relief vent. If this type of transformer is installed in a poorly ventilated area, it must be furnished with some method of absorbing gases that may be generated by arcing inside the case.

Any askarel-insulated transformer rated over 35,000 volts must be installed in a vault.

OIL-INSULATED TRANSFORMERS INSTALLED INDOORS

Many transformers are cooled and insulated with a special insulating oil. The fire hazard potential of oil-insulated transformers is greater than that of askarel-insulated transformers. Therefore, the *Code*® requirements are more exacting for oil-insulated transformers. See *NEC*®, *Article 450.26*.

OIL-INSULATED TRANSFORMERS INSTALLED OUTDOORS

The *Code*® requires that combustible buildings, door and window openings, and fire escapes must be safeguarded from fires originating in oil-insulated transformers. Such protection may be provided by effective space separation or by erecting a fire-resistant barrier between the transformer bank and the areas requiring protection.

In addition, the *Code*® requires that some means be installed to contain and remove the transformer oil from a ruptured transformer tank. Such a precaution applies to a transformer installation adjacent to a building where an oil explosion can result in a fire hazard without this preventive measure. See Figures 25-5 and 25-6 for examples of oil-insulated transformers installed outdoors.

Figure 25-5 Oil-filled transformers with radiator cooling fins.

Figure 25-6 Oil-insulated transformer used at a substation.

PROVISIONS FOR TRANSFORMER VAULTS

The *Code*® regulations cover all essential details for vaults used for transformer installations, including the arrangement, construction, and ventilation of the vaults. See *NEC*®, *Article 450, Part III*.

AUTOTRANSFORMERS

Code® specifications are provided for the use of autotransformers for lighting circuits. Recall that an autotransformer does not have separate primary and secondary windings. It consists of only one winding on an iron core. Part of the single winding of the autotransformer is common to both the primary and secondary circuits.

The *Code*® limits the use of an autotransformer when feeding branch circuits because of the interconnection of the primary and secondary windings. The autotransformer may be used only where the identified ground wire of the load circuit is connected solidly to the grounded wire of the source (*NEC*®, *210.9*). An exception to this rule is when a circuit without grounded conductors is used.

Figure 25-7 illustrates an autotransformer connected to a lighting load. Note that the grounded wire is carried through the entire system.

An alternate use of an autotransformer for lighting circuits is shown in Figure 25-8. This circuit also follows *Code*® regulations as the identified ground wire is carried through the entire system.

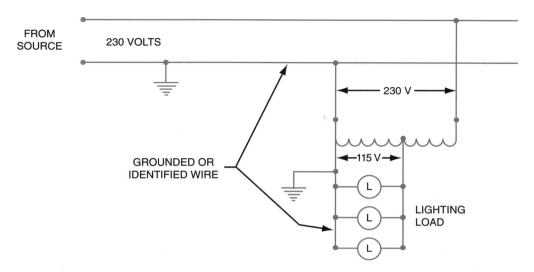

Figure 25-7 Approved use of an autotransformer.

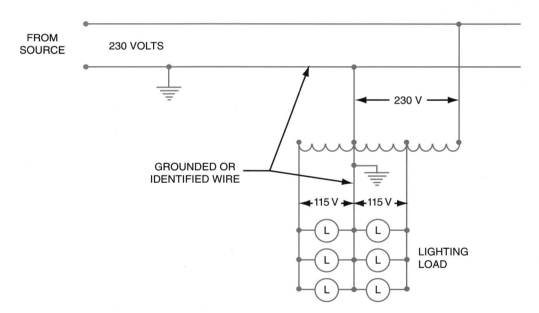

Figure 25-8 Approved use of an autotransformer.

Figure 25-9 illustrates an application for an autotransformer that is *not* approved by the *Code*®. Here, the single-phase, 230-volt input to the autotransformer is obtained from a three-phase, 230-volt source. The use of a mid-tap on the autotransformer makes available a single-phase, three-wire system for a lighting load. However, if the grounded wire is not solidly connected through the entire system, this circuit will not meet *Code*®

regulations. The type of installation shown in Figure 25-9
is unsafe, particularly if an
accidental ground condition
develops on the three-phase
system.

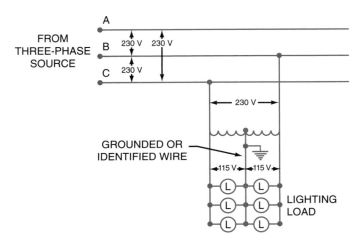

"Buck" and "Boost" Transformers

"Buck" and "boost"
transformers are forms of
autotransformer connections.
Transformers field connected
as an autotransformer are covered in *Article 450.4B*.

Figure 25-9 Incorrect use of an autotransformer.

The boost transformer is used to increase the supply voltage slightly. A typical application of a boost transformer connection is to increase the line voltage from a 208-V
source to be used with a 230-V rated machine. To do this, the line voltage only needs to
be increased 22 V. Rather than try to find an isolation-type transformer with a 208-V primary and a 230-V secondary, a simple transformer designed for boosting the voltage
22 volts is used. The boost transformer, or buck transformer, is a smaller, more economical solution than a full-size, fully rated primary and secondary. Typical values of added
voltage (boost) or subtracted voltage (buck) are secondaries of 12/24, 16/32, 24/48. The
primaries are generally dual-voltage rated as are the secondaries. By connecting the transformer as in Figure 25-10(A), the primary is rated for 120/240, but only 208 volts is available. Therefore, the primary voltage on each winding is one-half the total voltage or
104 V. This boost transformer is a 10:1 ratio (primary to secondary) so the secondary
windings rated for 12/24 V each develop 10.4 volts. The primary windings are in series
and the secondary windings are in series. The secondary voltage is added to the primary
voltage for a total of 228.8 volts or nearly the 320 V that was desired. The method of connecting a boost transformer is to connect X_1 to H_4. The supply lines are connected to H_1
and H_4, and the output connections are taken from H_1 and X_4. Notice that one line, H_1, is
common to both input and output, creating an autotransformer pattern.

Remember the instantaneous polarity of the H_1 lead is where current enters the primary winding or the negative polarity, and X_1 is where current leaves the secondary
winding at the same instant, the negative polarity. By assigning electrical polarities
(instantaneous values) to the leads, and connecting them as if they were batteries, you
can see that the voltages add together to yield an increased voltage.

To reduce the line voltage, you must "buck" the voltage. Use the same transformer
but reconnect the leads for bucking the applied voltage. Typically, this is used to reduce

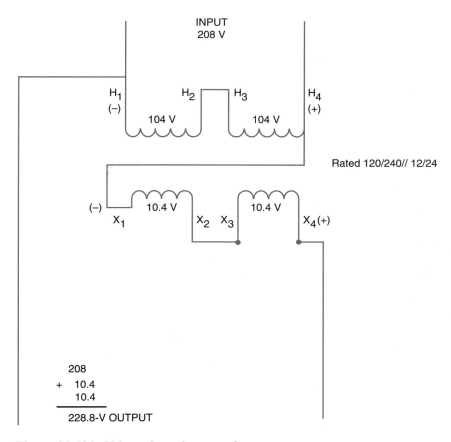

Figure 25-10A Voltage-boosting transformer.

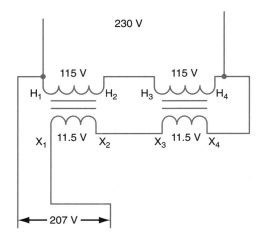

Figure 25-10B A standard connection for a "bucking" transformer connection used to lower voltage.

the voltage from 230 V to 208 V for equipment needs. The transformer are connected as in Figure 25-10(B) to have the secondary voltage oppose, or buck, the supply voltage. Notice that H_4 is connected to X_4. Again, if you assign instantaneous negative polarities to H_1 and X_1, you can see that the small secondary voltage is in opposition to the primary voltage. Subtract the secondary voltage from the primary voltage to get output values.

By using dual-voltage primaries and dual-voltage secondaries, you can make a variety of buck or boost voltages. Buck and boost transformers are rated in KVA. The load KVA determines the transformer used.

Figure 25-11 Buck or boost transformer's nameplates.

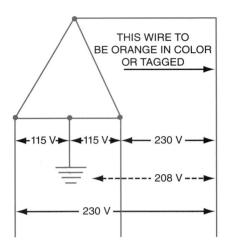

THIS WIRE TO BE ORANGE IN COLOR OR TAGGED

◄115 V► ◄115 V► ◄— 230 V —►

◄------ 208 V ---►

◄— 230 V —►

Figure 25-12 Four-wire delta secondary identification.

NEC® *Article 450.4(B)* states that a transformer field connected as an autotransformer, as is the case with boost transformers, shall be identified for use at the higher voltage. A transformer rated for boost or buck is designed for the higher voltage. *Article 210.9* refers to branch circuits derived from autotransformers and first states that the autotransformer shall be supplied with a grounded conductor. The exceptions to *Article 210.9* allow us to use the transformer as described above to increase or decrease voltage fed by two hot conductors.

Voltage drops (losses) in long or heavily loaded distribution systems may be increased in this manner to compensate for line drop. Many regular (isolating) transformer corrections are used as autotransformers to decrease or increase a voltage (see Figure 25-11).

IDENTIFIED FEEDER

On a four-wire, delta-connected secondary feeder conductor where the midpoint of one phase is grounded to supply lighting and similar loads, the phase conductor with the highest voltage to ground must be identified by an outer finish that is orange in color, or by tagging (see Figure 25-12). This identification is to be placed at any point where a connection is made if the neutral conductor is also present, such as in a distribution panel, junction box, or pull box. See *NEC*®, *Article 408.3(E)*.

SUMMARY

Transformers must be installed according to the provisions of the *NEC*®. There are many provisions that apply and each case must be thoroughly researched. When using the *Code*® criteria, be sure to consider all provisions of the *Code*®. Overcurrent protection can be quite different depending on the location and other protection available at the transformer installation. Be sure to consider the nameplate information when locating the transformer. Transformers can be dangerous and may cause severe injury or property damage if not installed and protected properly.

ACHIEVEMENT REVIEW

1. The rated primary current of a transformer is 4 amperes at 480 volts. How is the permissible maximum current setting determined for the overload devices used on the primary side of the transformer when secondary overcurrent protection is omitted?

    _____

    _____

    _____

    _____

    _____

2. If a transformer is rated less than 600 volts and has primary protection only, what percentage of the current is used for overcurrent protection if the primary current is 8 amps?

    a. _____

    _____

    b. _____

    _____

3. What items of data should appear on a transformer nameplate to comply with *Code*® requirements?

    _____

    _____

    _____

    _____

4. What is a secondary tie circuit? _____

_____

_____

_____

5. What happens to the secondary tie conductor size if loads are connected to the tie conductors?_____

_____

_____

_____

6. What overload devices are acceptable to protect a secondary tie connection with load taps? _____

_____

_____

_____

7. What are the electrical characteristics that must be similar if transformers are to be operated in parallel? _____

_____

_____

_____

8. What are the *Code*® requirements for grounding transformer installations?

_____

_____

_____

_____

9. How is the load divided when transformers are operated in parallel and protected as one unit?_____

_____

_____

_____

10. What precaution must be observed in using autotransformers to supply grounded lighting circuits?_____

    _____

    _____

    _____

    _____

11. Provide an example of how an autotransformer is used in a circuit to increase the voltage. _____

    _____

    _____

    _____

    _____

12. Connect the isolating transformer in Figure 25-13 as an autotransformer to "boost" the voltage from source to load. _____

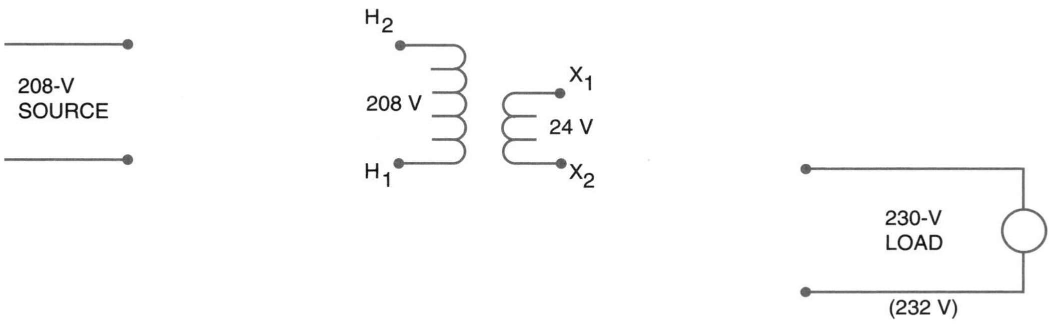

    _____

    _____

    _____

    _____

208-V
SOURCE

H_2

208 V

H_1

X_1

24 V

X_2

230-V
LOAD

(232 V)

Figure 25-13 Isolating transformer for question 12.

U•N•I•T
26
SUMMARY REVIEW
OF UNITS 17–25

OBJECTIVE

- To provide the student with an opportunity to evaluate the information and understanding acquired in the study of the previous 10 units.

1. What is a step-down transformer? _____

_____

_____

_____

_____

_____

_____

2. What is a step-up transformer? _____

_____

_____

_____

_____

_____

3. List the items that should be marked on the nameplate of a standard power or distribution transformer. _____

_____

_____

_____

_____

_____

4. Draw the schematic diagram of an additive polarity transformer.

5. Draw the schematic diagram of a subtractive polarity transformer.

For questions 6 through 17, select the correct answer for each of the statements and place the corresponding letter in the space provided.

6. The primary and secondary windings of an operating   _____
    transformer are tied together
    a. electrically.   c. through switching gear.
    b. magnetically.   d. not at all.

7. The H leads of a transformer are connected to the   _____
    a. high-voltage side.   c. secondary side.
    b. low-voltage side.   d. primary side.

8. The primary winding of a transformer is the   _____
    a. high-voltage side.   c. input winding.
    b. low-voltage side.   d. output winding.

9. The single-phase, three-wire system is   _____
    a. 115/230 volts.   c. 230 volts.
    b. 120/208 volts.   d. 120/277 volts.

10. An open-delta connection   _____
    a. is the same as a closed delta.
    b. is an incomplete connection.
    c. requires three single-phase transformers.
    d. requires two single-phase transformers.

11. The line voltage of a three-phase delta system is the same as   _____
    a. the line current.
    b. a single transformer voltage.
    c. a single transformer current.
    d. 1.73 × phase voltage.

12. The neutral of a three-phase, four-wire system is   _____
    a. grounded.   c. live.
    b. ungrounded.   d. bonded.

13. A V connection is the same as the   _____
    a. wye connection.   c. open-delta connection.
    b. delta connection.   d. open-wye connection.

14. Three 100-kVA transformers are connected in delta-delta.   _____
    The total kVA capacity is
    a. 58 percent of the three ratings.
    b. 58 percent of two ratings.
    c. 100 kVA.
    d. 300 kVA.

15. Five amperes is the standard rating of a(n)   _____
    a. instrument.
    b. secondary of a current transformer.
    c. secondary of a potential transformer.
    d. voltmeter movement.

16. It is dangerous to open the operating secondary of a   _____
    a. closed-delta transformer circuit.
    b. open-delta transformer circuit.
    c. potential transformer.
    d. current transformer.

17. A transformer with part of the primary serving as a   _____
    secondary is a(n)
    a. current transformer.   c. potential transformer.
    b. autotransformer.   d. open-delta transformer.

18. What are three standard types of cores used in transformers?

    a. _____

    b. _____

    c. _____

19. Name three common methods used to cool transformers.

    a. _____

    b. _____

    c. _____

20. A transformer has 1,200 turns in its primary winding and 120 turns in its secondary
    winding. The primary winding is rated at 2,400 volts. What is the voltage rating of
    the secondary winding? _____

21. State a simple rule to follow in connecting a transformer bank in closed delta.

    _____

    _____

    _____

22. What is one practical application of three single-phase transformers connected in a
    delta-delta configuration?

    _____

    _____

23. What is one practical application of an open-delta transformer bank?

    _____

    _____

    _____

    _____

24. The three-phase, four-wire secondary output of a wye-connected transformer bank can be used for the following two types of load:

    a. _____

    b. _____

25. State a simple rule that may be used in connecting single-phase transformers in wye.

    _____

    _____

    _____

26. What are two practical applications for a three-phase, delta-wye transformer bank?

    a. _____

    b. _____

27. What is one practical application for a three-phase, wye-delta transformer bank?

    _____

    _____

    _____

28. What is one advantage of using a three-phase transformer in place of three single-phase transformers? _____

    _____

    _____

29. What is one disadvantage of using a three-phase transformer in place of three single-phase transformers? _____

    _____

    _____

    _____

30. Insert the word or phrase to complete each of the following statements.
    a. A wye-delta transformer bank has _____-connected primary windings and _____-connected secondary windings.
    b. A delta-wye transformer bank has_____-connected primary windings and _____-connected secondary windings.
    c. A delta-wye transformer bank is used to _____ three-phase voltages to _____values.
    d. A wye-delta transformer bank is used to _____ extremely high three-phase voltages.
    e. A three-phase transformer takes _____ space than a transformer bank of the same kVA capacity consisting of three single-phase transformers.

31. List the five common three-phase connections used to connect transformer banks consisting of either two or three single-phase transformers.

    a. _____

    b. _____

    c. _____

    d. _____

    e. _____

32. What are the two distinct types of instrument transformers?

    a. _____

    b. _____

33. Why must the secondary circuit of a current transformer be closed when there is current in the primary circuit?

    _____

    _____

    _____

34. The transformers shown in Figure 26-1 have a 4 to 1 step-down ratio. Determine the secondary voltage of each transformer. In addition, determine the value that each voltmeter will indicate.

_____

_____

_____

_____

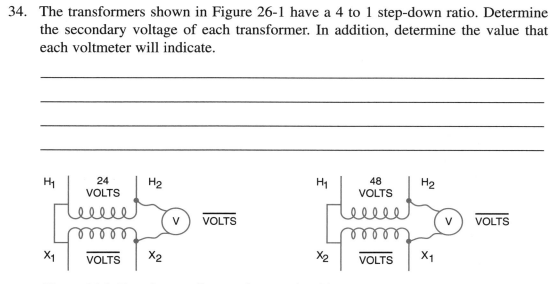

Figure 26-1 Transformer diagram for question 34.

35. Determine the unknown values. What is the polarity of the transformer in Figure 26-2?

(Additive, Subtractive)   _____

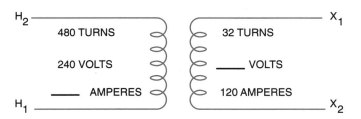

Figure 26-2 Transformer diagram for question 35.

36. Two 75-kVA transformers are connected in open delta. Determine the total kVA capacity of the transformers.

_____

_____

_____

_____

37. A transformer is marked 37.5 kVA. Its primary is rated at 480 volts and its second-
     ary is rated at 120 volts. Calculate the primary and secondary current ratings.

     (Primary)   _____

     (Secondary)   _____

     _____

     _____

     _____

     _____

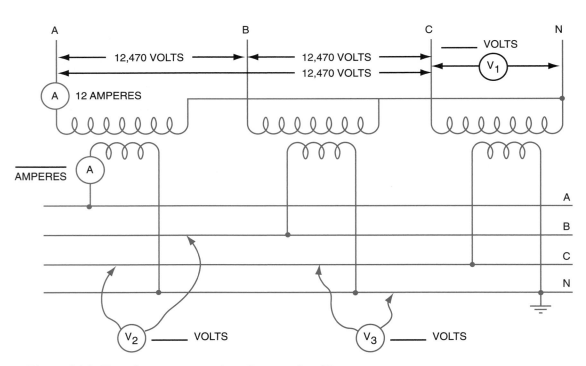

Figure 26-3 Transformer connections for question 38.

38. Calculate the values that will be indicated by the ammeter and the three voltmeters
     shown in Figure 26-3. The transformer ratio is 26 to 1. Insert the answers in the
     spaces provided on the diagram.

     _____

     _____

     _____

39. Explain the purpose of a "buck" or "boost" transformer. _____

    _____

    _____

    _____

    _____

40. On a four-wire delta system, what is an identified secondary feeder, and how is it identified? _____

    _____

    _____

    _____

    _____

    _____

    _____

GLOSSARY

ALTERNATING CURRENT (AC).   A current that alternates regularly in direction. Refers to a periodic current with successive half-waves of the same shape and area.

ARMATURE.   A cylindrical, laminated iron structure mounted on a drive shaft. It contains the armature winding.

ARMATURE WINDING.   Wiring embedded in slots on the surface of the armature. Voltage is induced in this winding on a generator.

AUTOTRANSFORMER.   A transformer in which a part of the winding is common to both the primary and secondary circuits.

BRUSH POLARITY.   Used to distinguish between the electrical polarity of the brushes and the magnetic polarity of the field poles.

BUCK OR BOOST TRANSFORMERS.   Transformers used to boost (increase) a voltage or to buck (lower) it. These are small amounts of change.

BUSWAY.   A system of enclosed power transmission that is current and voltage rated.

CIRCUIT BREAKER.   A device designed to open and close a circuit by nonautomatic means and to open the circuit automatically on a predetermined overcurrent without damage to itself when properly applied within its rating.

COMMUTATING POLES.   Interpoles energized by windings placed in series with the load circuit of a DC motor or generator.

COMMUTATOR.   Consists of a series of copper segments that are insulated from one another and the mounting shaft; used on DC motors and generators.

COMPOUND-WOUND GENERATOR.   A DC generator with a shunt and series, double-field winding.

CONDUIT PLAN.   A diagram of all external wiring between isolated panels and electrical equipment.

CONSTANT-CURRENT TRANSFORMERS.   Used for series street lighting where the current must be held constant with a varying voltage.

CORE-TYPE TRANSFORMER.   The primary is wound on one leg of the transformer iron, and the secondary is wound on the other leg.

COUNTER EMF.   An induced voltage developed in a DC motor while rotating. The direction of the induced voltage is opposite to that of the applied voltage.

CUMULATIVE COMPOUND-WOUND GENERATOR OR MOTOR.   A series winding is connected to aid the shunt winding.

CURRENT.   The rate of flow of electrons measured in amperes.

CURRENT FLOW.   The flow of electrons.

DELTA CONNECTION.   A circuit formed by connecting three electrical devices in series to form a closed loop. Used in three-phase connections.

DIFFERENTIAL COMPOUND-WOUND GENERATOR.   A series winding is connected to oppose the shunt winding.

DIODE.   A two-element device that permits current to flow through it in only one direction.

DIRECT CURRENT (DC).   Current that does not reverse its direction of flow. It is a continuous nonvarying current in one direction.

DISCONNECTING SWITCH.   A switch intended to open a circuit only after the load has been thrown off by some other means; not intended to be opened under load.

DISTRIBUTION TRANSFORMER.   Usually oil filled and mounted on poles, in vaults, or in manholes.

DOUBLE-WOUND TRANSFORMER.   Has a primary and a secondary winding. These two windings are independently isolated and insulated from each other.

EDDY CURRENT.   Current induced into the core of a magnetic device. Causes part of the iron core losses in the form of heat.

EFFICIENCY.   The efficiency of all machinery is the ratio of the output to the input.

$$\frac{\text{output}}{\text{input}} = \text{efficiency}$$

FEEDER.   The circuit conductor between the service equipment or the switchboard of an isolated plant and the branch circuit overcurrent device.

FLUX.   Magnetic field; lines of force around a magnet.

FUSE.   An overcurrent protective device with a circuit-opening, fusible part that is heated and severed by the passage of overcurrent through it.

GENERATOR.   Machine that changes mechanical energy into electrical energy. It furnishes electrical energy only when driven at a definite speed by some form of prime mover.

GROUNDED.   Connected to earth or to some conducting body that serves in place of earth.

HERTZ (Hz).   The measurement of the number of cycles of an alternating current or voltage completed in one second.

HYSTERESIS.   Part of iron core losses.

IDENTIFIED CONDUCTOR (NEUTRAL).   A grounded conductor in an electrical system, identified with the code color white.

INDUCED CURRENT.   Current produced in a conductor by the cutting action of a magnetic field.

INDUCED VOLTAGE.   Voltage created in a conductor when the conductor interacts with a magnetic field.

INDUCTION.   Induced voltage is always in such a direction as to oppose the force producing it.

INSTRUMENT TRANSFORMERS.   Used for metering and control of electrical energy, such as potential and current transformers.

INSULATOR.   Material with a very high resistance used to electrically isolate two conductive surfaces.

ISOLATING TRANSFORMER.   A transformer in which the secondary winding is electrically isolated from the primary winding.

LENZ'S LAW.   A voltage is induced in a coil whenever the coil circuit is opened or closed.

MOTORIZING.   A generator armature rotates as a motor.

NEC®.   National Electrical Code®.

OPEN DELTA.   Two transformers connected in a "V" supplying a three-phase system.

PARALLEL CIRCUIT.   A circuit that has more than one path for current flow.

PERMEABILITY.   The ease with which a material will conduct magnetic lines of force.

POLARITY.   Characteristic (negative or positive) of a charge. The characteristic of a device that exhibits opposite quantities, such as positive and negative, within itself.

POLE.   The north or south magnetic end of a magnet; a terminal of a switch; one set of contacts for one circuit of main power.

POLYPHASE.   An electrical system with the proper combination of two or more single-phase systems.

POLYPHASE ALTERNATOR.   A polyphase, synchronous alternating current generator, as distinguished from a single-phase alternator.

POWER FACTOR.   The ratio of true power to apparent power. A power factor of 100 percent is the best electrical system.

RATING.   The rating of a switch or circuit breaker includes (1) the maximum current and voltage of the circuit on which it is intended to operate, (2) the normal frequency of the current, and (3) the interrupting tolerance of the device.

RECTIFIER.   A device that converts alternating current (AC) into direct current (DC).

REGULATION.   Voltage at the terminals of a generator or transformer for different values of the load current; usually expressed as a percentage.

REMOTE CONTROL.   Controls the function initiation or change of an electrical device from some remote place or location.

RESIDUAL FLUX.   A small amount of magnetic field.

RHEOSTAT.   A resistor that can be adjusted to vary its resistance without opening the circuit in which it may be connected.

SEMICONDUCTOR.   Materials that are neither good conductors nor good insulators. Certain combinations of these materials allow current to flow in one direction but not in the opposite direction.

SEPARATELY EXCITED FIELD.   The electrical power required by the field circuit of a DC generator may be supplied from a separate or outside DC supply.

SERIES FIELD.   In a DC motor, has comparatively few turns of wire of a size that will permit it to carry the full-load current of the motor.

SERIES WINDING.   Generator winding connected in series with the armature and load carries full load.

SHELL-TYPE TRANSFORMER (DOUBLE WINDOW).   The primary and secondary coils are wound on the center iron core leg.

SHIELDED-WINDING TRANSFORMER.   Designed with a metallic shield between the primary and secondary windings; provides a safety factor by grounding.

SHORT AND GROUND.   A flexible cable with clamps on both ends. It is used to ground and short high lines to prevent electrical shock to workers.

SHUNT.   To connect in parallel; to divert or be diverted by a shunt.

SHUNT GENERATOR.   DC generator with its field connected in parallel with the armature and load.

SILICON-CONTROLLED RECTIFIER (SCR).   A four-layer semiconductor device that is a rectifier. It must be triggered by a pulse applied to the gate before it will conduct electricity.

SINGLE PHASE.   A term characterizing a circuit energized by a single alternating EMF. Such a circuit is usually supplied through two wires.

SOLID STATE.   As used in electrical-electronic circuits, refers to the use of solid materials as opposed to gases, as in an electron tube. It usually refers to equipment using semiconductors.

SPEED REGULATION.   Refers to the changes in speed produced by changes within the motor due to a load applied to the shaft.

STEP-DOWN TRANSFORMER.   With reference to the primary winding, the secondary voltage is lower.

STEP-UP TRANSFORMER.   The secondary voltage is higher than the primary voltage.

THREE PHASE.   A term applied to three alternating currents or voltages of the same frequency, type of wave, and amplitude. The currents and/or voltages are one-third of a cycle (120 electrical time degrees) apart.

THREE-PHASE SYSTEM.   Electrical energy originates from an alternator which has three main windings placed 120° apart. Three wires are used to transmit the energy.

TORQUE.   The rotating force of a motor shaft produced by the interaction of the magnetic fields of the armature and the field poles.

TRANSFORMER.   An electromagnetic device that converts voltages for use in power transmission and operation of control devices.

TRANSFORMER BANK.   When two or three transformers are used to step down or step up voltage on a three-phase system.

TRANSFORMER PRIMARY TAPS.   Alternative terminals that can be connected to more closely match the primary voltage supply.

TRANSFORMER PRIMARY WINDING.   The coil that receives the energy.

TRANSFORMER SECONDARY WINDING.   The coil that discharges the energy at a transformed or changed voltage, whether up or down.

UNDERCOMPOUNDING.   A small number of series turns on a compound DC generator that produces a reduced voltage at full load.

VOLTAGE CONTROL.   Intentional changes in the terminal voltage made by manual or automatic regulating equipment, such as a field rheostat.

WELDING TRANSFORMERS.   Provides very low voltages and high current to arc welding electrodes.

WIRING DIAGRAM.   Locates the wiring on a control panel in relationship to the actual location of the equipment and terminals. Specific lines and symbols represent components and wiring.

WYE CONNECTION (STAR).   A connection of three components made in such a manner that one end of each component is connected. This connection generally connects devices to a three-phase power system.

INDEX